GALLIUM ARSENIDE LASERS

Gallium Arsenide Lasers

Edited by

C. H. GOOCH
Services Electronics Research Laboratory, Baldock, Hertfordshire

WILEY - INTERSCIENCE

a division of John Wiley & Sons Ltd

London New York Sydney Toronto

Library of Congress catalog card number 78-93559

SBN 471 31320 3

PHYSICS

Printed in Great Britain by John Wright & Sons Ltd.,
at the Stonebridge Press, Bristol

Preface

The first semiconductor lasers were constructed towards the end of 1962. It was soon realized that these devices posed interesting theoretical and experimental problems and that they had many potential applications. The first devices were made in gallium arsenide and were excited using a $p-n$ junction structure. Soon other materials were being used and alternative methods of excitation, using electron beams, were being explored. However, it appears that gallium arsenide $p-n$ junction devices are by far the most important form of semiconductor laser and this book is concerned exclusively with this system.

Early devices had to be operated at low temperatures and under pulsed conditions and this limited their usefulness in applications. Consequently considerable effort has been devoted to achieving either continuous operation or operation at room temperature. Although continuous operation at room temperature has not been achieved devices now operate usefully at room temperature and continuously with some cooling.

This book gives a detailed account of various aspects of gallium arsenide lasers ranging from the preparation and properties of gallium arsenide to the applications of the devices. Included are chapters which give accounts of the theory and properties of devices and the technology involved in their fabrication.

It is thus hoped that the book will appeal to the diverse interests of theoretical and experimental semiconductor physicists and those concerned with the applications of semiconductor devices.

C. H. Gooch

Acknowledgements

Acknowledgement is made to the following sources for permission to reproduce copyright material.

Figs. 2.6, 2.7, 4.28. The American Institute of Physics.

Fig. 3.23. The North Holland Publishing Co.

Fig. 4.3. *Z. Metallkunde.*

Figs. 4.4, 6.12. Pergamon Press Inc.

Figs. 4.11, 4.12, 5.2. The Institute of Physics and The Physical Society.

Fig. 4.31. *Zeitschrift für Naturforschung.*

Figs. 5.19, 6.5, 6.6, 6.14, 6.20, 6.21. The Institute of Electrical and Electronic Engineers.

Chapters 3, 4 and 7 are published by permission of the Ministry of Defence (Navy).

Contributing Authors

M. J. Adams	*Department of Applied Mathematics and Mathematical Physics, University College, Cardiff*
R. F. Broom	*Institute of Applied Physics, University of Berne*
C. D. Dobson	*Standard Telecommunications Laboratories Ltd., Harlow, Essex*
C. H. Gooch	*Services Electronics Research Laboratory, Baldock, Hertfordshire*
K. G. Hambleton	*Services Electronics Research Laboratory, Baldock, Hertfordshire*
C. Hilsum	*Royal Radar Establishment, Malvern, England*
P. T. Landsberg	*Department of Applied Mathematics and Mathematical Physics, University College, Cardiff*
M. C. Rowland	*Services Electronics Research Laboratory, Baldock, Hertfordshire*

Contents

1

Introduction

C. HILSUM

Royal Radar Establishment, Malvern, England

It has often been said that the laser was a solution in search of a problem. It could also be said that for many years gallium arsenide was a semiconductor in search of an application. This introduction tells how these two searches coalesced, and opened up a new field.

The story really begins over forty years ago, when a number of the III–V compounds were first made, but it was not until the early 1950's that Russian and German workers independently suggested that this family of materials might form a useful addition to the few semiconductors of industrial importance (Boltaks and Zhuze, 1948, 1952; Welker, 1952). Most attention was concentrated on InSb, which could be prepared easily, and had an extraordinarily high electron mobility. At first sight GaAs also presented an attractive picture. It seemed that it might prove superior as a transistor material to either silicon or germanium, having both a large energy gap and a high electron mobility. However, these charms were not to fall like ripe fruit into the first hands that groped for them.

The problems that presented themselves in those early days were serious. The synthesis of GaAs was fraught with difficulty and even danger. Gallium, an innocuous metal at room temperature, is highly reactive at the melting point of GaAs, and at this temperature arsenic is a poisonous explosive vapour. During the first years of research it was common for the synthesis to be accompanied by an explosion. Though it was usually possible to pick pieces of GaAs from the wreckage, it was generally accepted that this method of preparation was inconvenient.

Progress was gradually made on developing a more controllable manufacturing technique, but even by 1960 the purity of the crystals available was still very low. A crude device technology had by then been established but the reproducibility and reliability were poor. At the same time the performance of devices made in silicon was improving rapidly. It was becoming apparent that GaAs could never compete on equal terms with silicon—an application was needed for which silicon was quite unsuitable. For a short period it seemed that tunnel diodes made in GaAs

1

would prove excellent fast switches, but soon it was discovered that these diodes were subject to a mysterious degradation. This was the low point for those working on GaAs. Logic pointed clearly to the abandonment of the field, and several laboratories reduced their efforts considerably. Others were unwilling to recognize this logic, and with obstinacy or insight, depending on one's point of view, carried on research. Now, however, there was a subtle difference in the approach. Previously the emphasis had been on GaAs as a conventional, albeit a superior, semiconductor. Now the tendency was to highlight ways in which GaAs differed not just in degree, but in kind. In this context attention became focused on the nature of the transition between conduction and valence bands.

It was recognized early that in GaAs the minimum energy transition was direct, requiring no cooperation with the lattice, whereas in silicon and germanium the same transition was indirect, involving a momentum exchange. Calculations of recombination probability had shown that recombination accompanied by the emission of radiation was far more likely for direct transitions than for indirect ones. This knowledge had been applied to InSb as early as 1955, when it was found that radiative recombination was a dominant process at low temperature, and was still important at room temperature. In the same year Braunstein (1955) made the first observations of radiation emission from GaAs.

It seems curious now that seven years lapsed after this with almost no further work on luminescence in GaAs. There were several theoretical papers pointing out that heavy doping was needed to make radiative recombination an important process in high energy gap materials, but there seemed little realization of the potentialities of semiconductor lamps, and little interest in the direct conversion of electricity into light. Pankove, early in 1962, appears to have been the first to appreciate that a GaAs p–n junction would prove an efficient source of infrared radiation (Pankove, 1962; Pankove and Berkeyheiser, 1962).

At this time lasers were well-established devices. The ruby laser was first made in 1960 and the gas laser in 1961. The possibility of obtaining coherent emission from semiconductors had been examined by Basov and coworkers (1959), before any laser had been made, and in 1961 Bernard and Duraffourg (1961) and Basov, Krokhin and Popov (1961) deduced that in a forward biased p–n junction, conditions could be favourable for obtaining an inverted population.

In 1962 the stage was set for the race to the first semiconductor laser. At Leningrad, in January of that year, Nasledov and coworkers (1962) reported that the line-width of the radiation emitted from GaAs diodes narrowed slightly at high current densities. He suggested that this might be a sign of stimulated emission, and announced his intention of trying specially shaped diodes to magnify these effects. Interest grew rapidly as summer approached, and in June Keyes and Quist (1962) announced they had constructed GaAs diodes with an internal quantum efficiency estimated as 85%. This encouraged more laboratories to work in the field, and the pace of the research increased. It was amid considerable excitement that Hall and coworkers (1962) at the General Electric Research Laboratories in Schenectady succeeded in making the first semiconductor laser towards the end of September.

A rival team led by Nathan at the IBM Research Laboratory was successful only ten days later, and in fact both reports were published on the same day (Hall and coworkers, 1962; Nathan and coworkers, 1962).

The heady atmosphere in which the semiconductor laser was created misled many workers on GaAs. They were sure that this discovery at once justified their years of fruitless research. It was certainly true that the resulting publicity drew many who had previously remained aloof to join in the rush to work on GaAs, and a great number of learned papers on lasers were published in the following two or three years. But as the excitement died down, and those who had clambered aboard the bandwagon jumped off, a more sober assessment became possible. The contribution made by the laser was not the promise of large device sales and commercial profits. It was much more that for the first time research on GaAs became respectable. Managements which had thought that 'semiconductor' was a synonym for 'silicon' realized that there were other fields to explore. Industrial and government financial backing increased, and a soundly based material and device technology was developed.

This was a period of consolidation, during which the tunnel diode fiasco was forgotten. The device which benefited most from the less prejudiced climate was not the laser, but the incoherent light emitting diode. The laser operated best at low temperature, and most people were not prepared to accept this complication. Naturally, much of the increased knowledge of luminescence and injection processes acquired during the research on diodes could be and was applied to give improved lasers, and this helped us to reach our present stage of development.

This is not to say that there are still no important applications for the laser. It is no part of an introduction to pre-empt the contents of the book, and to anticipate the material to come. Since a later section deals with laser applications, here we will only try to summarize future prospects. The laser remains a compact, highly efficient, fast converter of electricity into light. Its development is immature, and we are only gradually becoming aware of the phenomena which can be observed in more complex structures. The attainable power output increases steadily, and combination of individual diodes into arrays becomes more skilful. The low temperature of operation remains a disadvantage, but progress here too is good. Pulsed room temperature operation is now commonplace, and continuous operation has been achieved above 200°K.

Future semiconductor lasers will be both smaller and larger than they are today. The smaller devices will work continuously without cooling, the larger ones will give pulse powers measured in kilowatts. Multi-terminal devices will accomplish complicated switching operations, and will also produce radiation modulated at microwave frequencies. Progress on GaAs lasers will lead to improved lasers emitting visible radiation.

This book has not been written because the authors feel GaAs laser development is now complete. The first five years have seen us gaining a general understanding of the complex physical processes involved in this simple device. There is much physics we still have to learn, and much technology we have to master before we can fully apply the physics we have learned. But the foundations have been set. Later developments are much more likely to be based on a deeper appreciation of what we know today than on a realization that our theories are inadequate and misleading.

This is why it is now appropriate to set out our present knowledge on the gallium arsenide laser, in the belief that it will prove eventually one of the most important semiconductor devices.

References

Basov, N. G., B. Vul and Y. M. Popov (1959). *Zh. Eksperim. Teor. Fiz.*, **37**, 587 (*Soviet Phys. JETP*, **10**, 416, 1960).
Basov, N. G., O. N. Krokhin and Y. M. Popov (1961). *Zh. Eksperim. Teor. Fiz.*, **40**, 1879 (*Soviet Phys. JETP*, **13**, 1320, 1961).
Bernard, M. G. and G. Duraffourg (1961). *Phys. Stat. Solidi*, **1**, 699.
Boltaks, B. I. and V. P. Zhuze (1948). *Zh. Tekhn. Fiz.*, **18**, 1459.
Boltaks, B. I. and V. P. Zhuze (1952). *Izv. Akad. Nauk.*, **16**, 155.
Braunstein, R. (1955). *Phys. Rev.*, **99**, 1892.
Hall, R. N., G. E. Fenner, J. D. Kingsley, T. J. Soltys and R. O. Carlson (1962). *Phys. Rev. Letters*, **9**, 366.
Keyes, R. J. and T. M. Quist (1962). *Proc. I.R.E.*, **50**, 1822.
Nasledov, D. N., A. A. Rogachev, S. M. Ryvkin and B. V. Tsarenkov (1962). *Fiz. Tverd. Tela.*, **4**, 1062 (*Soviet Phys. Solid State*, **4**, 782, 1962).
Nathan, M. I., W. P. Dumke, G. Burns, F. H. Dill and G. J. Lasher (1962). *Appl. Phys. Letters.* **1**, 62.
Pankove, J. I. (1962). *Phys. Rev. Letters*, **9**, 283.
Pankove, J. I. and J. E. Berkeyheiser (1962). *Proc. I.R.E.*, **50**, 1976.
Welker, H. (1952). *Z. Naturforsch.*, **7a**, 744.

2

The Theory of the Injection Laser

M. J. ADAMS and P. T. LANDSBERG

Department of Applied Mathematics and Mathematical Physics,
University College, Cardiff

Contents

2.1   Introduction

The GaAs injection laser was first operated successfully in 1962 (Hall and coworkers, 1962; Nathan and coworkers, 1962; Quist and coworkers, 1962) following theoretical speculation (Dumke, 1962) as to the feasibility of obtaining coherent stimulated emission from semiconductors. Since then the basic theory has been established, although a complete under- standing of the device is hampered by the wide range of parameters which may affect its performance. Such quantities as doping levels and gradients, quantum efficiencies, absorption coefficients, temperature, geometrical structures and transport phenomena of carriers within the *p–n* junction all play a part in determining the properties of the laser.

This chapter seeks to give a self-contained account of the present theory of: (*a*) the radiative emission rates and their spectral dependence, and (*b*) the threshold current and its temperature dependence. The basic quantum mechanics and the statistical mechanics required for injection lasers is developed effectively *ab initio* in Sections 2.2 and 2.3. This treat- ment is, we believe, more thorough than usual. Readers not interested in the foundations may omit these sections on a first reading and refer to them only as the need arises in later sections. The model used for this develop- ment is that of a two-band semiconductor with or without a density-of- states 'tail' due to heavy doping. For the impurities, appropriate super- positions of Bloch functions are used. These are the essential ingredients which yield the emission rates (Sections 2.2 to 2.4). The approach is deficient in that it utilizes regions of constant quasi-Fermi levels, whereas, even if the radiative transitions occur mainly on the *p* side of the junction, this will not be correct. Thus carrier transport, including tunnelling, is neglected. However, these effects are corrections which are unlikely to affect the understanding gained by the simpler theory given here.

For a theory of the threshold current, however, the nature of the *p–n* junction enters in an essential way, since it provides the region in which the dominant recombination traffic takes place and also the waveguide which confines the radiation to the vicinity of this region. This part of the theory is given in Sections 2.5 and 2.6.

The presentation is perhaps novel in the way the matrix elements are derived in Section 2.4; in the comparison between the effects, particularly on threshold current, when the **k**-selection rule applies and the effects when it does not; in the discussion of the 'demerit' factor γ in Section 2.5; in the calculations with band tails in 2.5.6 and in the unified nature of the presentation. Apart from this no originality is claimed.

While Section 2.2 is quite general, and is needed for an understanding of lasers of any type, Sections 2.3 to 2.6 apply specifically to semiconductor lasers. In accordance with the spirit of this book we have not gone into other important theoretical problems which arise in connection with lasers generally, such as questions of coherence, energy partition between the lasing modes and transient effects.

No effort has been made to give exhaustive references to the literature for which the reader is referred to existing review articles (Burns and Nathan, 1964; Unger, 1965; Stern, 1966a; Nathan, 1966; Haken, 1965; Ivey, 1966; Popov, 1967; Pilkuhn, 1968).

We are grateful to Dr. K. Unger for his detailed comments on the manuscript.

2.2 Quantum mechanics of radiative transitions

The interaction of radiation and matter is usually treated by one of the following two methods:

[i] The semiclassical procedure which uses quantum mechanical perturbation theory for the electronic transition associated with the emission or absorption of radiation. It uses, however, a classical description of the radiation field. This merely enters through the electromagnetic vector potential **A** in the Hamiltonian

$$H_{\text{s.c.}} = \frac{1}{2m}\left[\mathbf{P} - \frac{e}{c}\mathbf{A}(\mathbf{r}, t)\right]^2 + e\phi(\mathbf{r}, t) + U(\mathbf{r}, t)$$

$$= \frac{1}{2m}P^2 - \frac{e}{2mc}(\mathbf{A}.\mathbf{P} + \mathbf{P}.\mathbf{A}) + e\phi + U \qquad (2.1)$$

where m is the mass of the electron, $\mathbf{P}$ its momentum operator, ϕ is the scalar electrostatic potential and U is the electron potential energy. In (2.1) the term in c^{-2} has been omitted. The theory yields the correct form of the absorption and stimulated emission probabilities. It does not, however, explain the occurrence of spontaneous emission. To understand this phenomenon in this theory, one has to rely on a variant of Einstein's original statistical argument (Einstein, 1917; Bohm, 1951).

[ii] In the correct treatment (Dirac, 1927; Bethe, 1964), the electromagnetic field is quantized and it occurs, as H_f say, in second-quantized form in the Hamiltonian

$$H = H_{\text{s.c.}} + H_f \tag{2.2}$$

of the system. This enables one to describe the field as a superposition of normal modes, or equivalent oscillators. A quantum-mechanical oscillator can emit radiation in the absence of an external field, and hence spontaneous emission occurs on the same footing as absorption or stimulated emission in this theory.

In this section an outline of approach [ii] will be given.

2.2.1   First-order time-dependent perturbation theory

Consider a quantum-mechanical system with Hamiltonian

$$H = H_0 + \lambda V(\mathbf{r}, t) \tag{2.3}$$

where H_0 is the unperturbed Hamiltonian, and λ is a constant which measures the strength of the perturbation. The (unperturbed) eigenfunctions of H_0 have the form[1]

$$\psi_I(\mathbf{r}) e^{-iE_I^0 t/\hbar}, \qquad H_0 \psi_I = E_I^0 \psi_I, \tag{2.4}$$

where the E_I^0 are the eigenvalues, and the ψ_I depends on all coordinates (briefly denoted by $\mathbf{r}$) specifying the system but not on the time t. Since the functions (2.4) form a complete orthonormal set, an eigenfunction ϕ of H can be expanded

$$\phi = \sum_R C_R(t) \psi_R e^{-iE_R^0 t/\hbar}. \tag{2.5}$$

Since ϕ must also satisfy the Schrödinger equation

$$i\hbar \frac{\partial \phi}{\partial t} = H\phi, \tag{2.6}$$

---

[1] The Dirac bra–ket notation will be used later. As it does not simplify the argument in this subsection, it is avoided here.

(2.5) must be inserted into (2.6) to yield

$$\sum_R (i\hbar \dot{C}_R + E_R^0 C_R) \psi_R e^{-iE_R^0 t/\hbar}$$
$$= \sum_R E_R^0 e^{-iE_R^0 t/\hbar} \psi_R C_R + \lambda \sum_R V C_R \psi_R e^{-iE_R^0 t/\hbar}.$$

Multiplying by $\psi_I^* \exp(iE_I^0 t/\hbar)$, integrating over all $\mathbf{r}$ and using the orthonormality of the ψ's,

$$\dot{C}_I = -\frac{i}{\hbar} \lambda \sum_R e^{i(E_I^0 - E_R^0)t/\hbar} V_{IR} C_R, \tag{2.7}$$

where

$$V_{IR} \equiv \int \psi_I^* V \psi_R \, d\mathbf{r} \tag{2.8}$$

can depend only on t.

If the unperturbed state is $R = J$ (say), and occurs for $t \leqslant t_0$ before the perturbation is 'switched on', then for short times later only the term $R = J$ will contribute in (2.7) when $I \neq J$. All other terms will involve higher corrections. Hence

$$C_I = -\frac{i}{\hbar} \lambda \int_{t_0}^t e^{i(E_I^0 - E_J^0)t/\hbar} V_{IJ} dt \qquad (I \neq J). \tag{2.9}$$

Now use (2.7) with $I = J$, and insert (2.9). To first order in λ,

$$\dot{C}_J = -\frac{i}{\hbar} \lambda V_{JJ} C_J, \quad \text{i.e.} \quad C_J = \exp\left[-\frac{i}{\hbar} \lambda \int_{t_0}^t V_{JJ} dt\right]. \tag{2.10}$$

Thus one has the expansion coefficients in terms of V.

Suppose now we have a periodic disturbance of angular frequency ω, as would happen for an electromagnetic wave. A Hermitian perturbation has then the form (the dagger denotes the Hermitian conjugate)

$$\lambda V = \sum_{\mathbf{k}} [G^{(\mathbf{k})} e^{-i\omega_{\mathbf{k}} t} + G^{(\mathbf{k})\dagger} e^{i\omega_{\mathbf{k}} t}], \tag{2.11}$$

where G is an operator which can involve the coordinates $\mathbf{r}$ as well as the momenta, but not the time. G and ω can depend on the label (wave vectors) $\mathbf{k}$. The time integration can then be carried out in (2.9), and one finds

$$-C_I = \sum_{\mathbf{k}} [G_{IJ}^{(\mathbf{k})} F_{IJ}(\omega_{\mathbf{k}}) + (G^{(\mathbf{k})\dagger})_{IJ} F_{IJ}(-\omega_{\mathbf{k}})] \qquad (I \neq J),$$

where the matrix element of G is defined as in (2.8) and

$$F_{IJ}(\omega_{\mathbf{k}}) \equiv \frac{1 - \exp\left[\frac{i}{\hbar}(t_0 - t)(E_I^0 - E_J^0 - \hbar\omega_{\mathbf{k}})\right]}{E_I^0 - E_J^0 - \hbar\omega_{\mathbf{k}}} \exp\left[\frac{it}{\hbar}(E_I^0 - E_J^0 - \hbar\omega_{\mathbf{k}})\right].$$
$$\tag{2.12}$$

The probability that a state $I \neq J$ will be found at time $t > t_0$ is given by

$$|C_I|^2 = \sum_k \{|G_{IJ}^{(k)} F_{IJ}(\omega_k)|^2 + |(G^{(k)\dagger})_{IJ} F_{IJ}(-\omega_k)|^2 + ...\}. \quad (2.13)$$

The dots denote terms for which the exponential factors in (2.12) do not cancel. These rapidly fluctuating terms do not contribute significantly. The two terms shown contribute only if $E_I^0 - E_J^0 \pm \hbar\omega_k$ can be small. After purely algebraic manipulations which involve the replacement of

$$\frac{\sin^2 y}{y^2} \quad \text{by} \quad \pi\delta(y) \qquad (2.14)$$

where $\delta(y)$ is the Dirac delta function, one finds for the transition probability per unit time

$$T_{IJ} = \frac{1}{t - t_0}|C_I(t)|^2 = \frac{2\pi}{\hbar} \sum_k \left[|G_{IJ}^{(k)}|^2 \delta(E_I^0 - E_J^0 - \hbar\omega_k) \right.$$

$$\left. + |G_{IJ}^{(k)\dagger}|^2 \delta(E_I^0 - E_J^0 + \hbar\omega_k) \right] \qquad (I \neq J). \qquad (2.15)$$

This result is basic to the later work. The first delta function shows that a state of higher energy can be reached by the part of the system which does not include the radiation if a quantum of the appropriate frequency is available for absorption. The sum must be retained in (2.15) since a number of wave vectors $\mathbf{k}$ may satisfy the delta-function condition.

2.2.2 Fourier analysis of the pure radiation field

In order to obtain the Hamiltonian H_f of a free electromagnetic field, it is desirable to avoid the continuous infinity of generalized coordinates and momenta needed to describe it. This set is made enumerable by assuming the field to be enclosed in a cube of side L, and to require that the values of the vector potential $\mathbf{A}$ and its derivatives have the same values at pairs of opposite points on the surface. This is the familiar periodic boundary condition. The permitted wave vectors $\mathbf{k} = (k_1, k_2, k_3)$ for $\mathbf{A}$ form then a discrete set:

$$k_j = \frac{2\pi n_j}{L} \qquad (j = 1, 2, 3 \quad n_j = 0, \pm 1, \pm 2, ...). \qquad (2.16)$$

(Periodic boundary condition)

One can now make a spatial Fourier expansion of $\mathbf{A}$ for any time t, in the form

$$\mathbf{A}(\mathbf{r}, t) = \sum_{\mathbf{k}, \lambda} q_{\mathbf{k}\lambda}(t) \mathbf{A}_{\mathbf{k}\lambda}(\mathbf{r}). \qquad (2.17)$$

Here $\lambda = 1, 2, 3$ for the three mutually perpendicular directions of polarization. Let $e_{k\lambda}$ be a triad of orthonormal real unit vectors for each k (associated with the directions of polarization). One can then, in accordance with usual Fourier expansions, interpret the $A_{k\lambda}$ as plane waves (in both the classical and the quantum treatments):

$$\mathbf{A}_{k\lambda} = \frac{a_k}{\sqrt{V}} e^{ik \cdot r} \mathbf{e}_{k\lambda}, \tag{2.18}$$

so that $\mathbf{A}_{k\lambda}^* = \mathbf{A}_{-k\lambda}$, $a_k = a_{-k}$, $\mathbf{e}_{k\lambda} = \mathbf{e}_{-k\lambda}$, where a_k is a real normalization constant and $V = L^3$. The k sum in (2.17) goes over the whole of k space. Since A is real (classically) or a Hermitian operator (quantum mechanically) one has from (2.17) and (2.18)

$$\mathbf{A} - \mathbf{A}^\dagger = 0 = \sum_{k,\lambda} (q_{k\lambda} \mathbf{A}_{k\lambda} - q_{k\lambda}^\dagger \mathbf{A}_{k\lambda}^*) = \sum_{k,\lambda} (q_{k\lambda} - q_{-k\lambda}^\dagger) \mathbf{A}_{k\lambda}.$$

While the asterisk indicates the complex conjugate of an ordinary function, the dagger can here serve the double purpose of indicating the complex conjugate in the classical theory and a Hermitian conjugate operator in its quantum-mechanical analogue. The reality condition therefore is

$$q_{k\lambda}^\dagger = q_{-k\lambda}$$

so that

$$\dot{q}_{k\lambda}^\dagger = \dot{q}_{-k\lambda}. \tag{2.17'}$$

By (2.18) the $A_{k\lambda}$ are orthogonal:

$$\int_V \mathbf{A}_{k\lambda}^* \cdot \mathbf{A}_{k'\lambda'} \, d\mathbf{r} = a_k^2 \, \delta_{k,k'} \, \delta_{\lambda,\lambda'} \quad \text{or} \quad \int_V \mathbf{A}_{k\lambda} \cdot \mathbf{A}_{k'\lambda'} \, d\mathbf{r} = a_k^2 \, \delta_{k',-k} \, \delta_{\lambda,\lambda'}. \tag{2.19}$$

The magnetic field is

$$\mathbf{H} = \text{curl } \mathbf{A} = i \sum_{k,\lambda} q_{k\lambda} (\mathbf{k} \times \mathbf{A}_{k\lambda}) \tag{2.20}$$

For longitudinally polarized waves let us choose $\lambda = 3$. Then

$$\mathbf{e}_{k3} = \mathbf{k}/|\mathbf{k}|. \tag{2.21}$$

Then (2.20) shows that these waves do not contribute to the magnetic field, and one can omit the value $\lambda = 3$ in (2.20). By (2.18) this convention also means that $\text{div } \mathbf{A}_{k\lambda}$ is proportional to $\mathbf{k} \cdot \mathbf{e}_{k\lambda}$ and so vanishes for $\lambda = 1, 2$. Hence for the transverse part of the field

$$\mathbf{A}_{tr} \equiv \sum_{\substack{k \\ \lambda=1,2}} q_{k\lambda} \mathbf{A}_{k\lambda}, \quad \text{div } \mathbf{A}_{tr} = 0. \tag{2.22}$$

It is convenient to split the electric field into transverse and longitudinal components. These satisfy

$$\mathbf{E}_{\mathrm{tr}} = -\frac{1}{c}\frac{\partial \mathbf{A}_{\mathrm{tr}}}{\partial t}, \quad \mathbf{E}_{\mathrm{lg}} = -\frac{1}{c}\frac{\partial \mathbf{A}_{\mathrm{lg}}}{\partial t} - \nabla\phi,$$

ϕ being the scalar potential, and together they yield Maxwell's equation

$$\mathbf{E} = \mathbf{E}_{\mathrm{tr}} + \mathbf{E}_{\mathrm{lg}} = -\frac{1}{c}\frac{\partial \mathbf{A}}{\partial t} - \nabla\phi. \tag{2.23}$$

From (2.17) and (2.22)

$$\mathbf{E}_{\mathrm{tr}} = -\frac{1}{c}\sum_{k,\lambda=1,2}\dot{q}_{k\lambda}\mathbf{A}_{k\lambda}. \tag{2.24}$$

A free field, sometimes called a pure radiation field, such as occurs for light waves, does not involve electric charges and longitudinal components are absent (Heitler, 1954). For such a field the energy is

$$W = \frac{1}{8\pi}\int (H^2 + E_{\mathrm{tr}}^2)\,d\mathbf{r} = \frac{1}{8\pi}\int_V (\mathbf{H}^\dagger.\mathbf{H} + \mathbf{E}^\dagger.\mathbf{E})\,d\mathbf{r}. \tag{2.25}$$

The second form is more convenient for the quantum-mechanical case. This quantity will be evaluated in order to determine restrictions on a_k.

The first integral is, if $\mathbf{k}\times\mathbf{e}_{k1} = |\mathbf{k}|\,\mathbf{e}_{k2}$ and $\mathbf{k}\times\mathbf{e}_{k2} = -|\mathbf{k}|\,\mathbf{e}_{k1}$,

$$\frac{1}{8\pi}\sum_{\substack{kk'\\\lambda\lambda'=1,2}}\int q_{k'\lambda'}^\dagger q_{k\lambda}(\mathbf{k}'\times\mathbf{A}_{k'\lambda'}^*).(\mathbf{k}\times\mathbf{A}_{k\lambda})\,d\mathbf{r}$$

$$= \frac{1}{8\pi}\sum_{kk'\lambda}q_{k'\lambda}^\dagger q_{k'\lambda}|\mathbf{k}|\,|\mathbf{k}'|\int_V \mathbf{A}_{k'\lambda}^*.\mathbf{A}_{k\lambda}\,d\mathbf{r} \tag{2.26}$$

$$= \frac{1}{8\pi}\sum_{k,\lambda=1,2}a_k^2|\mathbf{k}|^2 q_{k\lambda}^\dagger q_{k\lambda}$$

The second integral is

$$\frac{1}{8\pi c^2}\sum_{k,\lambda=1,2}a_k^2\dot{q}_{k\lambda}^\dagger\dot{q}_{k\lambda} \tag{2.27}$$

This shows that the integrals are easily evaluated.

2.2.3 The free field as a collection of oscillators

Classically (2.25) yields

$$W = \frac{1}{8\pi c^2}\sum_{\substack{k\\\lambda=1,2}}a_k^2(\dot{q}_{k\lambda}^\dagger\dot{q}_{k\lambda} + |kc|^2 q_{k\lambda}^\dagger q_{k\lambda}) \tag{2.28}$$

The energy of a classical harmonic oscillator of mass m and angular frequency ω is

$$\frac{m}{2}(\dot{q}^2 + \omega^2 q^2) = \frac{m}{2}\left[\left(\frac{p}{m}\right)^2 + \omega^2 q^2\right] \qquad (2.29)$$

This suggests that the energy of the field can be thought of as residing in an enumerable infinity of non-interacting one-dimensional harmonic oscillators with angular frequency.

$$\omega_{\mathbf{k}} = \omega_{-\mathbf{k}} = |\mathbf{k}|c \qquad (2.30)$$

To test this idea we try a Hamiltonian $H_{\mathrm{f}}(p_1, p_2, ..., q_1, q_2, ...)$ for the transverse part of the field suggested by (2.28) and (2.29):

$$\left.\begin{aligned} H_{\mathrm{f}} &= \frac{1}{2}\sum_{\substack{\mathbf{k}\\ \lambda=1,2}} A_{\mathbf{k}}\,[B_{\mathbf{k}}^2 p_{\mathbf{k}\lambda}^{\dagger} p_{\mathbf{k}\lambda} + \omega_{\mathbf{k}}^2 q_{\mathbf{k}\lambda}^{\dagger} q_{\mathbf{k}\lambda}], \\[2mm] q_{\mathbf{k}\lambda}^{\dagger} &= q_{-\mathbf{k}\lambda}, \quad p_{\mathbf{k}\lambda}^{\dagger} = p_{-\mathbf{k}\lambda} \end{aligned}\right\} \qquad (2.31)$$

where $A_{\mathbf{k}} = A_{-\mathbf{k}}$ is real and has the dimension of mass, $B = B_{-\mathbf{k}}$ and $A_{\mathbf{k}} B_{\mathbf{k}}$ is a number. In fact, from (2.28)

$$A_{\mathbf{k}} \equiv \frac{\mu^2 a_{\mathbf{k}}^2}{4\pi c^2}, \qquad \dot{q}_{\mathbf{k}\lambda} \equiv B_{\mathbf{k}} p_{\mathbf{k}\lambda}^{\dagger}, \qquad (2.32)$$

where we have allowed for a refractive index μ of the (non-magnetic) medium containing the radiation.

Treating (2.31) classically, one can use the Hamiltonian equations of motion

$$\dot{q}_{\mathbf{k}\lambda} = \frac{\partial H_{\mathrm{f}}}{\partial p_{\mathbf{k}\lambda}} = A_{\mathbf{k}} B_{\mathbf{k}}^2 p_{\mathbf{k}\lambda}^{\dagger}, \quad \dot{p}_{\mathbf{k}\lambda} = -\frac{\partial H_{\mathrm{f}}}{\partial q_{\mathbf{k}\lambda}} = -A_{\mathbf{k}}\omega_{\mathbf{k}}^2 q_{\mathbf{k}\lambda}^{\dagger} \qquad (2.33)$$

By (2.32, 2.33) one can eliminate $B_{\mathbf{k}}$ since it must satisfy

$$A_{\mathbf{k}} B_{\mathbf{k}} = 1$$

for consistency. Relations (2.33) also imply

$$\ddot{q}_{\mathbf{k}\lambda} = B_{\mathbf{k}}\dot{p}_{\mathbf{k}\lambda}^{\dagger} = -\omega_{\mathbf{k}}^2 q_{\mathbf{k}\lambda} \qquad (2.34)$$

confirming that (2.31) is an oscillator Hamiltonian.

Quantum mechanically, one treats the p's and q's in (2.31) as operators subject to the commutation rules

$$\left.\begin{aligned} i[p_{\mathbf{k}\lambda}, q_{\mathbf{k}'\lambda'}] &= \hbar\delta_{\mathbf{k},\mathbf{k}'}\,\delta_{\lambda,\lambda'} \\[2mm] i[p_{\mathbf{k}\lambda}, p_{\mathbf{k}'\lambda'}] &= i[q_{\mathbf{k}\lambda}, q_{\mathbf{k}'\lambda'}] = 0. \end{aligned}\right\} \qquad (2.35)$$

By the Heisenberg equations of motion,

$$\left.\begin{array}{l} \dot{q}_{k\lambda} = \dfrac{i}{\hbar}[H_f, q_{k\lambda}] = p^\dagger_{k\lambda}/A_k \\[3mm] \dot{p}_{k\lambda} = \dfrac{i}{\hbar}[H_f, p_{k\lambda}] = -A_k\,\omega^2_k q^\dagger_{k\lambda} \end{array}\right\} \tag{2.36}$$

Also, using (2.36),

$$\ddot{q}_{k\lambda} = \frac{i}{\hbar}[H_f, \dot{q}_{k\lambda}] = \frac{i}{A_k\hbar}[H_f, p^\dagger_{k\lambda}] = \frac{1}{A_k}\dot{p}^\dagger_{k\lambda} = -\omega^2_k q_{k\lambda} \tag{2.37}$$

This is the exact analogue of the classical result embodied in (2.33) and (2.34). In fact (2.34) and (2.37) suggest that dimensionless quantities $b_{k\lambda}, b'_{k\lambda}$ exist such that

$$q_{k\lambda} = [b_{k\lambda}\,e^{-i\omega k l} + b'^\dagger_{k\lambda}\,e^{i\omega k l}]\,Q_k, \quad Q_k = Q_{-k} = \text{real.} \tag{2.38}$$

Equation (2.17′) enables one to eliminate the b' by virtue of

$$b'_{k\lambda} = b_{-k\lambda} \tag{2.39}$$

Q_k, like A_k and B_k above, carries the appropriate dimension. Classically the b's are numbers, quantum mechanically they are operators.

If one applies (2.17′) to (2.33) or (2.36) one also finds an analogous result for the p's:

$$p^\dagger_{k\lambda} = p_{-k\lambda} \tag{2.40}$$

We now interpret the operators b in terms of the generalized coordinates and momenta and hence determine their commutation relations. By (2.36) and (2.38),

$$p^\dagger_{k\lambda} = -i\omega_k A_k Q_k[b_{k\lambda}\,e^{-i\omega k l} - b^\dagger_{-k\lambda}\,e^{i\omega k l}] \tag{2.41}$$

Eliminating $b_{-k\lambda}$ from (2.38) and (2.41),

$$b_{k\lambda} = \frac{1}{2Q_k}\left[q_{k\lambda} + \frac{i}{\omega_k A_k}p^\dagger_{k\lambda}\right]e^{i\omega k l} \tag{2.42}$$

An analogous relation for $b^\dagger_{-k\lambda}$ is obtainable similarly, but is in fact already implied by (2.42). From (2.35) and (2.42) one now finds

$$[b_{k\lambda}, b^\dagger_{k'\lambda'}] = \frac{\hbar}{2\omega_k A_k Q^2_k}\delta_{k,k'}\,\delta_{\lambda,\lambda'} \tag{2.43}$$

$$[b_{k\lambda}, b_{k'\lambda'}] = [b^\dagger_{k\lambda}, b^\dagger_{k'\lambda'}] = 0$$

By virtue of (2.38) and (2.41) the Hamiltonian H_f of (2.31) can be expressed entirely in terms of the operators b. Utilizing (2.43) one finds

$$H_f = \sum_{\substack{k \\ \lambda=1,2}} \{(2\omega_k A_k Q_k^2/\hbar) b_{k\lambda}^\dagger b_{k\lambda} + \tfrac{1}{2}\} \hbar\omega_k \qquad (2.44)$$

It is convenient to define the Q_k now by

$$2\omega_k A_k Q_k^2 = \hbar, \quad \text{i.e.} \quad a_k Q_k = \sqrt{(2\pi\hbar/\omega_k)}\,(c/\mu) \qquad (2.45)$$

so that finally

$$H_f = \sum_{\substack{k \\ \lambda=1,2}} (b_{k\lambda}^\dagger b_{k\lambda} + \tfrac{1}{2}) \hbar\omega_k, \qquad (2.46)$$

where

$$[b_{k\lambda}, b_{k'\lambda'}^\dagger] = \delta_{k,k'} \delta_{\lambda,\lambda'} \qquad (2.47)$$

2.2.4 The number operator

The operator $N_{k\lambda} \equiv b_{k\lambda}^\dagger b_{k\lambda}$ satisfies by virtue of (2.43) and (2.46)

$$N_{k\lambda} b_{k\lambda} = (b_{k\lambda} b_{k\lambda}^\dagger - 1) b_{k\lambda} = b_{k\lambda}(N_{k\lambda} - 1) \qquad (2.48)$$

Taking Hermitian conjugates

$$N_{k\lambda} b_{k\lambda}^\dagger = b_{k\lambda}^\dagger (N_{k\lambda} + 1) \qquad (2.49)$$

it follows that: (a) if $|N_{k\lambda}'\rangle$ is the eigenfunction of $N_{k\lambda}$ with eigenvalue $N_{k\lambda}'$, then $b_{k\lambda}|N_{k\lambda}'\rangle$ is the eigenfunction of $N_{k\lambda}$ with eigenvalue $N_{k\lambda}' - 1$ and $b_{k\lambda}^\dagger|N_{k\lambda}'\rangle$ is the eigenfunction of $N_{k\lambda}$ with eigenvalue $N_{k\lambda}' + 1$; (b) $N_{k\lambda}' \geqslant 0$ because $N_{k\lambda}' = \langle N_{k\lambda}'|N_{k\lambda}|N_{k\lambda}'\rangle = \langle b_{k\lambda} N_{k\lambda}'|b_{k\lambda} N_{k\lambda}'\rangle$.

The last expression is the length of a vector and hence non-negative. By repeated application of $b_{k\lambda}$ as in (a) one reaches eventually negative eigenvalues, contrary to (b), unless the eigenvalue zero stops the process in accordance with (2.48). Hence the eigenvalue zero of $N_{k\lambda}$ occurs, and the other eigenvalues are the positive integers. The operator $N_{k\lambda}$ is therefore called the number operator. Its eigenvalue gives by (2.46) the occupation number of the mode $(\mathbf{k}, \lambda)$.

All the $|N_{k\lambda}'\rangle$ eigenfunctions are assumed normalized. Hence

$$\langle b_{k\lambda}^\dagger N_{k\lambda}'|b_{k\lambda}^\dagger N_{k\lambda}'\rangle = \langle N_{k\lambda}'|b_{k\lambda} b_{k\lambda}^\dagger|N_{k\lambda}'\rangle = \langle N_{k\lambda}'|N_{k\lambda} + 1|N_{k\lambda}'\rangle = N_{k\lambda}' + 1$$

$$\langle b_{k\lambda} N_{k\lambda}'|b_{k\lambda} N_{k\lambda}'\rangle = N_{k\lambda}'.$$

It follows that

$$\left. \begin{aligned} |b_{k\lambda}^\dagger N_{k\lambda}'\rangle &\equiv b_{k\lambda}^\dagger|N_{k\lambda}'\rangle = \sqrt{N_{k\lambda}' + 1}\,|N_{k\lambda}' + 1\rangle \\ |b_{k\lambda} N_{k\lambda}'\rangle &\equiv b_{k\lambda}|N_{k\lambda}'\rangle = \sqrt{N_{k\lambda}'}\,|N_{k\lambda}' - 1\rangle \end{aligned} \right\} \qquad (2.50)$$

Finally, one sees from (2.50) that

$$|N'_{k\lambda}\rangle = [N'_{k\lambda}!]^{-\frac{1}{2}} [b^\dagger_{k\lambda}]^{N'_{k\lambda}} |0\rangle \qquad (2.51)$$

$$\left.\begin{array}{l} \langle N''_{k\lambda'}|b^\dagger_{k\lambda}|N'_{k\lambda}\rangle = \sqrt{N'_{k\lambda}+1}\ \delta_{N''_{k\lambda},\,N'_{k\lambda}+1}\ \delta_{\lambda,\lambda'} \\[2mm] \langle N''_{k\lambda'}|b_{k\lambda}|N'_{k\lambda}\rangle = \sqrt{N'_{k\lambda}}\ \delta_{N''_{k\lambda},\,N'_{k\lambda}-1}\ \delta_{\lambda,\lambda'} \end{array}\right\} \qquad (2.52)$$

The $b^\dagger_{k\lambda}$ is called the creation operator for mode (k, λ), while $b_{k\lambda}$ is called the annihilation operator.

2.2.5 The perturbation

In order to find the form of the perturbation, the following observations are needed. The vector potential is by (2.17), (2.38) and (2.39)

$$\mathbf{A} = \sum_{\substack{k \\ \lambda=1,2}} [b_{k\lambda}\mathbf{A}_{k\lambda}e^{-i\omega kl} + b^\dagger_{-k\lambda}\mathbf{A}_{k\lambda}e^{i\omega kl}]Q_k$$

If in the second sum one replaces $-\mathbf{k}$ by $\mathbf{k}'$ and uses (2.18), (2.30) and (2.38) one finds

$$\mathbf{A} = \sum_{\substack{k \\ \lambda=1,2}} [b_{k\lambda}\mathbf{A}_{k\lambda}e^{-i\omega kl} + b^\dagger_{k\lambda}\mathbf{A}^*_{k\lambda}e^{i\omega kl}]Q_k \qquad (2.53)$$

Next note that for $j = 1,\ 2$ or 3

$$i(P_j\mathbf{A} - \mathbf{A}P_j) = \hbar\frac{\partial\mathbf{A}}{\partial x_j}$$

Multiplying by a unit vector parallel to x_j and summing,

$$i(\mathbf{P}.\mathbf{A} - \mathbf{A}.\mathbf{P}) = \hbar\,\mathrm{div}\,\mathbf{A} = 0 \qquad (2.54)$$

The last form holds because of (2.22).

The eigenfunctions of the unperturbed electron system are denoted by $I\rangle$, corresponding to the Hamiltonian $H_0 \equiv P^2/2m + e\phi + U$. The wave functions $|N'_{k\lambda}\rangle$ correspond to the free radiation field, specified by the Hamiltonian H_f of (2.46). The combined system has Hamiltonian (2.2), i.e.

$$H = H_0 + H_f + H_1, \quad H_1 \equiv -\frac{e}{mc}\mathbf{A}.\mathbf{P} \qquad (2.55)$$

where (2.54) has been used. Using (2.53) and (2.18) one finds

$$H_1 = \sum_k [G^{(k)}e^{-i\omega kl} + G^{(k)\dagger}e^{i\omega kl}] \qquad (2.56)$$

where

$$G^{(k)} \equiv -\frac{ea_k}{mc\sqrt{V}}Q_k \sum_{\lambda=1,2} b_{k\lambda}e^{ik.r}\mathbf{e}_{k\lambda}.\mathbf{P}$$

Thus the perturbation has the form (2.11) and the transition probability is given by (2.15). Using (2.45) it requires the calculation of a matrix element of

$$G^{(k)} = -\sqrt{\left(\frac{2\pi\hbar}{V\omega_k}\right)} \frac{e}{\mu m} \sum_{\lambda=1,2} b_{k\lambda} e^{i\mathbf{k}\cdot\mathbf{r}} \mathbf{e}_{k\lambda}\cdot\mathbf{P} \tag{2.57}$$

2.2.6 The transition probability

The unperturbed problem corresponds to product wave functions

$$\prod_{\mathbf{k}} \prod_{\lambda=1}^{2} \{|N'_{k\lambda}\rangle\}$$

for the field and $|I\rangle$ for the electron system. As a result of the transition to be considered only one radiation mode gains or loses a photon. By (2.57) and (2.52) the modes which are left unchanged do not contribute to the matrix element. Let the mode for which there is a change be the $(\mathbf{k}, \lambda)$ mode; the remaining modes can then be neglected. Let the initial state be $|I\rangle|N'_{k\lambda}\rangle$ and let the electron system make a transition to the state $|J\rangle$. Let the final state of the radiation mode $(\mathbf{k}, \lambda)$ be $|N''_{k\lambda}\rangle$. Then by (2.52) a non-zero matrix element is obtained only if $N''_{k\lambda} = N'_{k\lambda}+1$, when the contribution comes from $G^\dagger$, and $E^0_J < E^0_I$ (emission). Alternatively $N''_{k\lambda} = N'_{k\lambda}-1$, when the contribution comes from G and $E^0_J > E^0_I$ (absorption). Hence the matrix element is

$$\left.\begin{matrix}|G^{(k)\dagger}_{IJ}|\\ |G^{(k)}_{IJ}|\end{matrix}\right\} = -\sqrt{\frac{2\pi\hbar}{V\omega_k}}\frac{e}{\mu m}\cdot\begin{cases}\displaystyle\sum_{\lambda=1,2} M^-_{IJ}\sqrt{N'_{k\lambda}+1} & \text{(emission)}\\[2ex] \displaystyle\sum_{\lambda=1,2} M^+_{IJ}\sqrt{N'_{k\lambda}} & \text{(absorption)}\end{cases}$$

where

$$M^\pm_{IJ} \equiv M^\pm_{IJ}(\mathbf{k},\lambda) \equiv |\langle J|e^{\pm i\mathbf{k}\cdot\mathbf{r}}\mathbf{e}_{k\lambda}\cdot\mathbf{P}|I\rangle| \tag{2.58}$$

Substitution in (2.15) yields for the transition probability per unit time

$$T_{IJ} = \frac{4\pi^2 e^2}{\mu^2 V m^2} \sum_{\substack{\mathbf{k}\\ \lambda=1,2}} \frac{1}{\omega_k} \begin{cases}(M^-_{IJ})^2(N'_{k\lambda}+1)\,\delta(E^0_I - E^0_J - \hbar\omega_k)\\ (M^+_{IJ})^2 N'_{k\lambda}\,\delta(E^0_I - E^0_J + \hbar\omega_k)\end{cases} \tag{2.59}$$

Assuming isotropic radiation in a medium of refractive index μ, so that $E_k = \dfrac{\hbar k c}{\mu}$, the $\mathbf{k}$ sum in (2.59) may be converted into integrals over the photon energy E_k and the solid angle $d\Omega_k$ due to the possible directions of **k**. The density-of-states function required is

$$\frac{V}{8\pi^3} d\mathbf{k} = \frac{Vk^2\, dk\, d\Omega_\mathbf{k}}{8\pi^3} = \frac{V\mu^3\, E_\mathbf{k}^2}{8\pi^3 \hbar^3 c^3} dE_\mathbf{k}\, d\Omega_\mathbf{k} = \mathcal{N}(E_\mathbf{k})\, dE_\mathbf{k} \frac{d\Omega_\mathbf{k}}{4\pi} \quad (2.60)$$

where $\mathcal{N}(E_\mathbf{k})$ is the number of modes of given polarization per unit energy. Hence,

$$T_{IJ} = \frac{4e^2\,\mu}{m^2\,\hbar^2\,c^3} \int \frac{E_\mathbf{k}^2}{\hbar\omega_\mathbf{k}} \left\{ \begin{array}{l} (N'_{\mathbf{k}\lambda}+1)\,\delta(E_I^0 - E_J^0 - \hbar\omega_\mathbf{k}) \\ N'_{\mathbf{k}\lambda}\,\delta(E_I^0 - E_J^0 + \hbar\omega_\mathbf{k}) \end{array} \right\} \left[\sum_{\lambda=1,2} \int (M_{IJ}^\pm)^2 \frac{d\Omega_\mathbf{k}}{8\pi} \right] dE_\mathbf{k}$$

The polarization vectors $\mathbf{e}_{\mathbf{k}\lambda}$ which enter the matrix element M_{IJ} have unit length and depend in direction only on λ and on the *direction* of **k**. We can therefore separate the above integrations by adopting the following assumptions:

[i] Since the photon wave vector **k** is generally small, the term $e^{\pm i\mathbf{k}\cdot\mathbf{r}}$ in the matrix element will be neglected.

[ii] The occupation number $N'_{\mathbf{k}\lambda}$ of mode $(\mathbf{k}, \lambda)$ will be assumed to be close to its equilibrium value so that $N'_{\mathbf{k}1} = N'_{\mathbf{k}2}$ depends only on $|\mathbf{k}|$, i.e. on $E_\mathbf{k}$. This puts the λ sum in the correct place in the above expression.

The term in square brackets can therefore be integrated over a solid angle of magnitude 4π, and it then represents the average (over the directions of **k** and the values of λ) of $(M_{IJ})^2$. It will be denoted by $M_{IJ}|_{\mathrm{av}}^2$. The $E_\mathbf{k}$-integration leads to

$$T_{IJ} = \frac{4e^2\,E\mu}{m^2\,\hbar^2\,c^3} |M_{IJ}|_{\mathrm{av}}^2 \left\{ \begin{array}{ll} (N_\nu+1) & \text{(emission)} \\ (N_\nu) & \text{(absorption)} \end{array} \right. \quad (2.61)$$

where the subscript ν indicates a photon energy $E \equiv h\nu \equiv E_I^0 - E_J^0$.

The *absorption* probability is proportional to N_ν and there is an emission term also proportional to N_ν. This is called the *stimulated* emission probability. There remains a term independent of N_ν and this is called the *spontaneous emission* probability, since it does not require the presence of a field.

2.3   Recombination statistics[2]

2.3.1   *Probabilities for quantum states*

The preceding section was quantum-mechanical, but included the statistics of the radiation field in (2.60). In this section the statistics of the electron system is developed. To this end consider a transition of this

---

[2] See also Landsberg (1967a). A fuller account of recombination statistics is given in this paper together with references to the literature. This paper also contains a number of new results.

system from a state I to a state J, the latter having the lower energy. Disregarding the detailed nature of the process, we treat the many-body problem in a single-particle approximation, and stipulate that we are dealing with a single-electron transition. The states of the remaining electrons are assumed unaffected by the transition. The forward rate may then be written as

$$p_I S_{IJ} q_J \tag{2.62}$$

where p_I is the probability that the quantum state I is in a condition in which it can give up an electron, while q_J is the probability of finding a specific state which can be converted into the state J under consideration by the addition of an electron.

These definitions are best explained by two examples.

[i] *State I corresponds to an electron in a conduction band state at energy E_I.* In that case, writing $\eta_I = E_I/kT$

$$p_I = [1 + \exp(\eta_I - F_c)]^{-1} \tag{2.63}$$

where $\mu_c \equiv kTF_c$ is the quasi-Fermi level for the conduction band. The use of this concept implies that all conduction band electrons are assumed to be in equilibrium with each other, and are therefore subject to a Fermi–Dirac distribution. However, the quasi-chemical potential or quasi-Fermi level which occurs in it is not necessarily the same as for other groups of levels (valence bands, impurities, etc.).

In thermal equilibrium all quasi-Fermi levels in a given problem go over into a single Fermi level $\mu_0 \equiv kTF_0$.

Now q_I is in this case uniquely the probability of finding the same state unoccupied, so that the addition of an electron will correspond to state I. Hence

$$q_I = 1 - p_I \tag{2.64}$$

Note that

$$q_I/p_I = \exp(\eta_I - F_c) \tag{2.65}$$

The name 'probability coefficient' has been proposed for this ratio (Landsberg, 1967a); it is sufficiently important and simple to deserve a special name.

[ii] *State J corresponds to an r-electron impurity centre at position l being in its αth quantum state.* In this case it is most convenient to use the grand canonical distribution when in thermal equilibrium, denoted by the suffix o,

$$(p_J)_o \rightarrow (p_{r\alpha l})_o = \frac{\exp[-\eta(r, \alpha, l)]}{\displaystyle\sum_{s=0}^{M} \exp(sF_o) Z_{sl}} e^{rF_o} \tag{2.66}$$

Here α labels the ground and excited states of the r-electron centre, $kT\eta(r, \alpha, l)$ is the energy of the centre in this state, and Z_{sl} is the partition function for the centre when it has captured s electrons, i.e.

$$Z_{sl} = \sum_\alpha \exp\left[-\eta(s, \alpha, l)\right] \qquad (2.67)$$

The values o and M in (2.66) are somewhat arbitrary. An o-electron centre is one which has given up all the electrons it will ever give up in a certain context; an M-electron centre is one which has captured the maximum number (M) of electrons it is ever going to capture in the same context. In fact there is rarely any need to go beyond the values $M = 3$ or 4.

It is convenient and often useful (though not always correct) to assume that a quasi-Fermi level exists for each state of charge of the centre. We shall denote it by $kT(F_{1l} + F_{2l} + \ldots F_{rl})$ for an r-electron centre. The assumption involved here is that the transitions within a trap of given charge occur more rapidly than the transitions in which a trap changes its charge. The non-equilibrium steady-state generalization of (2.66) is then

$$p_J \rightarrow p_{r\alpha l} = \frac{\exp\left[F_{1l} + F_{2l} + \ldots + F_{rl} - \eta(r, \alpha, l)\right]}{Z_{ol} + \displaystyle\sum_{s=1}^{M} \exp\left(F_{1l} + F_{2l} + \ldots + F_{sl}\right) Z_{sl}} \qquad (2.68)$$

The assumption involved can be put in another way: all that is implied in (2.68) is the assumption

$$p_{r\alpha l} / p_{r\beta l} = e^{\eta(r,\beta,l) - \eta(r,\alpha,l)} \qquad \text{(all } r, l, \alpha, \beta) \qquad (2.69)$$

The F_{sl}'s can then be fixed in accordance with the total probabilities

$$p_{sl} = \sum_\alpha p_{s\alpha l}$$

which occur, but about whose relative magnitudes nothing has been assumed.

In many cases all centres are equivalent so that the quantities F_{sl}, $\eta(r, \alpha, l)$, $p_{r\alpha l}$ are the same for all positions l where impurities can be found. In that case the suffices l can of course be omitted, and this will be done here from the next subsection.

Turning to q_J, the 'specified state' referred to in the definition is, in contrast with example [i], not unique. Any of the ground or excited states of an $(r-1)$ centre will be acceptable. Choosing a particular β value, then, together with $r-1$ and l (about which there is no choice),

$$q_J \rightarrow p_{r-1\beta l} \qquad (2.70)$$

The probability coefficient is by (2.68) and (2.70)

$$\frac{q_J}{p_J} \rightarrow \frac{p_{r-1\beta l}}{p_{r\alpha l}} = \exp\left[\eta(r, \alpha, l) - \eta(r-1, \beta, l) - F_{rl}\right] \qquad (2.71)$$

Note that, in contrast with (2.64),

$$p_J + q_J \neq 1 \qquad (2.72)$$

It is because of the complications which can arise with impurities, that the meaning of p_I and q_J in (2.62) must be most carefully specified. Of course state I can be an impurity state in some problems, and state J can refer to a state in a band. In considering the consequences of (2.62) in this section we are not committed to any of the possible specific interpretations.

The analogy between the probability coefficient for a band and that for an r-electron centre is by (2.65) and (2.71)

$$\eta_I \leftrightarrow \eta(r, \alpha, l) - \eta(r-1, \beta, l) \qquad (2.73)$$

2.3.2 Recombination rate for single-electron transitions (use of detailed balance)

In fact one is rarely interested in the bare expression (2.62). First, the converse process is always going on, though it may be negligible in some cases. For the net recombination rate it should be subtracted from (2.62). Secondly, transitions like these occur always between groups of quantum states. So let us effect a summation of terms like (2.62) over all the states I which lie in a group, denoted by i, which all have the same quasi-Fermi level, to be denoted by kTF_i. Then the recombination rate is

$$u_{ij} = \sum_{\substack{I \in i \\ J \in j}} (p_I S_{IJ} q_J - p_J S_{JI} q_I) \qquad (2.74)$$

Each difference in (2.74) vanishes in thermal equilibrium. This is essentially the principle of detailed balance. It yields

$$\left(\frac{S_{JI}}{S_{IJ}}\right)_0 = \left(\frac{p_I\,q_J}{q_I\,p_J}\right)_0 = e^{\eta_J - \eta_I} \qquad (2.75)$$

If one of the states is localized, the replacement (2.73) is needed in the exponent. For *radiative transitions* (2.61) shows that if the distribution of radiation in the material, specified by the occupation numbers N_ν is never far from its equilibrium value, then one can put

$$\frac{S_{JI}}{S_{IJ}} = \left(\frac{S_{JI}}{S_{IJ}}\right)_0 \qquad (2.76)$$

2

We are not in this subsection concerned only with radiative transitions; (2.76) is in fact often a good approximation and it will be adopted in the present subsection but not in the remainder of Section 2.3. It is then convenient to define the following product of probability coefficients.

$$Z_{IJ} \equiv \frac{q_I \, p_J}{p_I \, q_J} = e^{\eta_I - F_i} e^{F_j - \eta_J} = e^{F_j - F_i + \eta_I - \eta_J} \tag{2.77}$$

From (2.76) the quantity

$$X_{IJ} \equiv S_{JI} Z_{IJ} / S_{IJ} = e^{F_j - F_i} \tag{2.77'}$$

depends only on the quasi-Fermi levels of the two groups of states involved in the transition, and not on the states themselves. Hence (2.74) becomes

$$u_{ij} = \sum_{IJ} p_I S_{IJ} q_J (1 - X_{IJ}) = p_{ij} (1 - e^{F_j - F_i}) \tag{2.78}$$

where

$$p_{ij} \equiv \sum_{I,J} p_I S_{IJ} q_J \tag{2.79}$$

is the total forward rate. The argument teaches that the ratio of reverse rate to forward rate is

$$\frac{\text{reverse rate}}{\text{forward rate}} = e^{F_j - F_i} \tag{2.80}$$

The net recombination rate vanishes in equilibrium when $F_j = F_i = F_0$. The summations in (2.78) are mainly of two types: (a) over the states in an almost empty conduction band or an almost full valence band; (b) over the localized states of identical r-electron centres. For the former there is a spread in the energies of the electron states. For the latter the energy spread is often negligible, so that the summation is equivalent to multiplying by the number involved.

In some cases it is important to allow for the fact that two-electron transitions, i.e. electron collisions, can contribute significantly to the rate u_{ij}. This generalization is possible but is omitted here. Subject to the omission of other transitions, the above exposition holds generally for single-electron transitions.

2.3.3 Radiative recombination rates

In this subsection two alternative arguments are offered for the discussion of the relation between stimulated and spontaneous emission.

[i] *An elementary argument based on the preceding subsection.* We treat as known the existence of stimulated and spontaneous emission rates and

of an absorption rate, given respectively in the forms

$$B_{IJ} N_\nu p_I q_J, \quad C_{IJ} p_I q_J, \quad B_{IJ} N_\nu p_J q_I$$

where B_{IJ} and C_{IJ} are independent of statistics, i.e. $B_{IJ} = (B_{IJ})_0$. As in (2.61) N_ν denotes the occupation number of a mode of frequency ν. The equality of the transition probabilities for induced emission and for absorption has been built into these formulae.

The relation between B_{IJ} and C_{IJ} is treated as unknown. One can then identify the transition probabilities of the preceding subsection as

$$S_{IJ} = B_{IJ} N_\nu + C_{IJ}, \quad S_{JI} = B_{IJ} N_\nu \qquad (2.81)$$

Equation (2.77') yields at once

$$(X_{IJ})_0 = \left\{ \frac{B_{IJ} N_\nu Z_{IJ}}{B_{IJ} N_\nu + C_{IJ}} \right\}_0 = 1, \quad \text{i.e.} \quad C_{IJ} = B_{IJ} \{N_\nu (Z_{IJ} - 1)\}_0 \qquad (2.82)$$

Also the ratio of the net stimulated emission rate to the spontaneous emission rate is given by

$$\frac{u_{IJ}^{\mathrm{st}} - u_{IJ}^{\mathrm{abs}}}{u_{IJ}^{\mathrm{sp}}} = \frac{B_{IJ}}{C_{IJ}} N_\nu (1 - Z_{IJ}) = \frac{N_\nu (1 - Z_{IJ})}{\{N_\nu (Z_{IJ} - 1)\}_0} \qquad (2.83)$$

We now observe that $\eta_I - \eta_J = h\nu/kT$ if the electronic transition leads to the emission of a photon of frequency $h\nu$. Also the Bose distribution yields

$$N_{\nu 0} = 1 \Big/ \left[\exp\!\left(\frac{h\nu}{kT}\right) - 1 \right] \qquad (2.84)$$

It follows, using (2.77), that Bose and Fermi distributions combine to yield

$$\{N_\nu (Z_{IJ} - 1)\}_0 = 1 \qquad (2.85)$$

With these interpretations the required relationships (2.82) and (2.83) take the forms

$$B_{IJ} = C_{IJ} \qquad (2.86)$$

$$\frac{1}{N_\nu} \frac{u_{IJ}^{\mathrm{st}} - u_{IJ}^{\mathrm{abs}}}{u_{IJ}^{\mathrm{sp}}} = 1 - e^{F_j - F_i + h\nu/kT} \qquad (2.87)$$

It is to be noted that the distribution N_ν can differ from the equilibrium one because (2.76) has in fact not been used in this subsection. Had it been used, (2.81) shows that the assumption $N_\nu \sim N_{\nu 0}$ would have been implied. While (2.77) holds for radiative transitions, equations (2.78) to (2.80) hold for radiative transitions only if $N_\nu \sim N_{\nu 0}$. It is therefore of interest to work out the ratio (2.80) for the present case. It is

$$\frac{\text{reverse rate}}{\text{forward rate}} = \frac{B_{IJ} N_\nu p_J q_I}{B_{IJ} (N_\nu + 1) p_I q_J} = \frac{N_\nu}{N_\nu + 1} Z_{IJ} = \frac{N_\nu}{N_\nu + 1} e^{F_j - F_i + \eta_I - \eta_J}$$

$$(2.87')$$

This reduces to (2.80) only if $N_\nu \sim N_{\nu 0}$ as given by (2.84).

The approach [i] involves a number of unproved assumptions in connection with (2.81), and does not lead to an evaluation of the quantities (2.86). These shortcomings are removed in the next approach.

[ii] *An argument based on* (2.61). From (2.61) one can write down immediately, using the probability factors introduced in this section,

$$\left.\begin{aligned}
u_{IJ}^{\text{st}} &= B_{IJ} N_\nu p_I q_J \\
u_{IJ}^{\text{abs}} &= B_{IJ} N_\nu p_J q_I \\
u_{IJ}^{\text{sp}} &= B_{IJ} p_I q_J
\end{aligned}\right\}$$

$$(2.88)$$

where

$$B_{IJ} \equiv \frac{4e^2 \mu E}{m^2 \hbar^2 c^3} |M_{IJ}|_{\text{av}}^2$$

$$(2.89)$$

Equations (2.86) and (2.87) follow at once.

The next step is to sum over the states $I \in i$ and $J \in j$ where i and j denote the groups of states specified by the Fermi levels $\mu_i \equiv kTF_i$ and $\mu_j = kTF_j$. Equation (2.78) contains the prototype of such summations. For this purpose regard the photon energy E absorbed or emitted as a variable and multiply the expressions (2.88) by the number of single-electron states $g_{IJ}(E) dE$ which can contribute to the emission of a photon in the energy range $(E, E + dE)$. Now I and J will range over a clearly defined set of states and each stands for a definite set of quantum numbers, I and J being linked by $\eta_I - \eta_J = E/kT$. Also the emission and absorption of radiation will be referred to an element $d\Omega$ of solid angle. One can then write for the recombination rates per unit volume

$$\frac{1}{V} \int \left\{ \sum_{IJ} (u_{IJ}^{\text{st}} - u_{IJ}^{\text{abs}}) g_{IJ}(E) \, \delta_{E_I - E_J, E} \right\} dE \frac{d\Omega_{\text{k}}}{8\pi}$$

$$\equiv \int \left\{ \sum_{IJ} r_{IJ}^{\text{st}} \right\} N_\nu \, dE \equiv \int r_{ij}^{\text{st}} N_\nu \, dE \qquad (2.90)$$

$$\frac{1}{V} \int \left\{ \sum_{IJ} u_{IJ}^{\text{sp}} g_{IJ}(E) \, \delta_{E_I - E_J, E} \right\} dE \frac{d\Omega_{\text{k}}}{8\pi}$$

$$\equiv \int \left\{ \sum_{IJ} r_{IJ}^{\text{sp}} \right\} dE \equiv \int r_{ij}^{\text{sp}} \, dE \equiv R_{ij}^{\text{sp}} \qquad (2.91)$$

From (2.88) and (2.89) one then has

$$r_{ij}^{\text{sp}} = \sum_{I,J} \frac{4\mu e^2 E |M_{IJ}(E)|_{\text{av}}^2}{V m^2 \hbar^2 c^3} g_{IJ}(E) p_I q_J \, \delta_{E_I - E_J, E} \qquad (2.92)$$

Hence, using (2.87)

$$r_{ij}^{\text{st}} = (1 - e^{F_j - F_i + E/kT}) r_{ij}^{\text{sp}} \qquad (2.93)$$

The dimensions of these quantities are (energy $\times$ volume $\times$ time)$^{-1}$.

2.3.4 Absorption coefficient

The net number rate of photon absorption per unit volume is

$$- \int r_{ij}^{\text{st}} \, dE,$$

where the minus sign changes emission to absorption.

The incident photon flux (number per unit area per unit time) is

$$\frac{2}{V} \int \frac{c}{\mu} \mathcal{N}(E) \, dE \left(\frac{d\Omega_k}{4\pi} \right)$$

where $\mathcal{N}(E)$ is given by (2.60). The ratio of these two quantities for any increment of energy E is defined to be the absorption coefficient α_{ji}. It follows that

$$\alpha_{ji} = - \frac{\pi^2 c^2 \hbar^3}{\mu^2 E^2} r_{ij}^{\text{st}} \qquad (2.94)$$

It can be related to r_{ij}^{sp} by means of (2.93).

2.3.5 The condition for negative absorption

This situation arises if for a given pair of states I, J

$$u_{IJ}^{\text{st}} > u_{IJ}^{\text{abs}}$$

The condition for stimulated emission to dominate absorption in this sense can be read off eqn. (2.87) to be (Bernard and Duraffourg, 1961)

$$F_i - F_j > h_\nu / kT \quad \text{or} \quad \mu_i - \mu_j > h\nu \qquad (2.95)$$

In thermal equilibrium the Gibbs free energy of various groups of $n_1, n_2, \ldots$ fermions having chemical potentials $\mu_1, \mu_2, \ldots$ is

$$G = \sum_r n_r \mu_r \qquad (2.96)$$

In a steady non-equilibrium state one uses quasi-chemical potentials (or quasi-Fermi levels) μ_r and (2.96) can then be regarded as the quasi-Gibbs free energy. If (2.96) applies before a radiative transition, then after the transition the quasi-Gibbs free energy is

$$G' = n_1 \mu_1 + \ldots + (n_i - 1) \mu_i + \ldots + (\mu_j + 1) \mu_j + \ldots$$

Let $\Delta G = G - G'$ be the drop in quasi-Gibbs free energy as a result of one transition in which a photon of frequency ν is emitted. Then (2.95) can be be given the different form (Landsberg, 1967b)

$$\Delta G > h\nu \qquad\qquad (2.97)$$

The form (2.97) of the condition remains valid for a wider range of physical situations than have been considered here, and is more general than (2.95).

2.3.6   *The condition for population inversion*

One may without reference to radiative processes enquire under what conditions a quantum state I of energy E_I has a greater occupation probability than a quantum state J of energy E_J lying below it. This is the phenomenon of population inversion. In accordance with our earlier notation, suppose that I belongs to a group of quantum states of quasi-Fermi level $\mu_i = kTF_i$, and J has quasi-Fermi level μ_j. The condition is clearly

$$\frac{1}{\exp(\eta_I - F_i) + 1} > \frac{1}{\exp(\eta_J - F_j) + 1}$$

It reduces to

$$F_i - F_j > (E_I - E_J)/kT \qquad\qquad (2.98)$$

The two conditions are clearly equivalent. By the analogy (2.73) this relation holds also if impurity states are involved.

2.4   Emission rates

In order to apply the general theory of Sections 2.2 and 2.3 to the semiconductor laser, it is necessary to: (*a*) identify the electronic transition(s) responsible for the radiative emission, (*b*) calculate a suitable matrix element, M_{IJ}, and (*c*) calculate the emission rates (involving a knowledge of densities-of-states functions and quasi-Fermi level positions, etc.).

We shall deal with these three problems in order.

2.4.1 Possible transitions

As a first approximation we consider the energy diagram (Fig. 2.1) of a direct-gap semiconductor which contains shallow donor and acceptor impurities. There are then four obvious possible radiative transitions which would give emission of approximately the wavelength found from semiconductor lasers. These are: (i) band–band, (ii) conduction band–acceptor, (iii) donor–valence band and (iv) donor–acceptor.

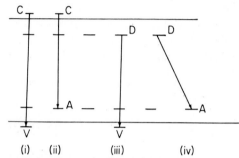

Fig. 2.1. The radiative transitions to be considered

We shall not deal here with the possibility of exciton transitions being involved (Knox, 1963; Wilson, 1963); nor are we concerned with indirect transitions (Landsberg, 1967c) since no lasing has been reliably reported from indirect gap materials.

The simplest approach to a quantitative description of the states involved in the transitions of Fig. 2.1 is to assume that the band states are represented by Bloch functions:

$$\psi_n(\mathbf{k}, \mathbf{r}) = V^{-\frac{1}{2}} u_{nk}(\mathbf{r})\, e^{i\mathbf{k} \cdot \mathbf{r}} \tag{2.99}$$

normalized over the volume of the crystal V. The shallow impurity states are represented by (Kohn, 1957)

$$\psi = \sum_{n,\mathbf{k}} u_{nk}(\mathbf{r})\, G_{nk} \tag{2.100}$$

where G_{nk} is the Fourier coefficient of a hydrogen-like wave function, into which an effective mass and the dielectric constant are introduced. The bands are labelled by the suffix n.

Fig. 2.1 is an idealized energy-band diagram of a semiconductor laser. There is usually a very high concentration of impurities which may then be thought of as forming impurity bands. These impurity bands can merge with the nearest band to form a conduction or valence band tail. Thus we may expect the simple wave functions (2.99), (2.100) to be greatly modified for a highly doped semiconductor, especially in the presence of band tails when the impurity wave functions will certainly not be very localized. Nevertheless, in the absence of more accurate expressions, (2.99) and (2.100) have been used as approximate wave functions. Rough estimates of the (averaged) matrix elements involved in the various transitions can thus be obtained, using (2.58) the photon wave vector.

2.4.2 Matrix elements

[i] *Band–band transitions* (Fig. 2.1(i)). Using the Bloch functions (2.99), denoted by c and v for conduction and valence band wave functions, respectively, the interband matrix element which is required is

$$\mathbf{P}_{vc}(\mathbf{k}_v, \mathbf{k}_c) = \frac{1}{V} \int_V u^*_{vk_v}(\mathbf{r})\, e^{-i\mathbf{k}_v \cdot \mathbf{r}}\, e^{-i\kappa \cdot \mathbf{r}}\, \mathbf{P} u_{ck_c}(\mathbf{r})\, e^{i\mathbf{k}_c \cdot \mathbf{r}}\, d\mathbf{r}$$

$$= \frac{1}{V} \int_V u^*_{vk_v}(\mathbf{r})\, e^{i(\mathbf{k}_c - \mathbf{k}_v - \kappa) \cdot \mathbf{r}} (\mathbf{P} + \hbar \mathbf{k}_c)\, u_{ck_c}(\mathbf{r})\, d\mathbf{r} \qquad (2.101)$$

where κ is the photon wave vector. This is of the form

$$I \equiv \int_V e^{i\lambda \cdot \mathbf{r}} f(\mathbf{r})\, d\mathbf{r} = N \sum_m \left\{ \delta_{\lambda, K_m}\!\left(\int_\Omega e^{iK_m \cdot \mathbf{r}} f(\mathbf{r})\, d\mathbf{r} \right) \right\}$$

where $\lambda \equiv \mathbf{k}_c - \mathbf{k}_v - \kappa$, Ω is the volume of the unit cell, and

$$f(\mathbf{r}) \equiv \frac{1}{V} u^*_{vk_v} (\mathbf{P} + \hbar \mathbf{k}_c)\, u_{ck_c}$$

is lattice periodic. The sum is over all reciprocal lattice vectors $\mathbf{K}_m$ and theorems for integrals of type I are known (Landsberg, 1969). Hence,

$$\mathbf{P}_{vc}(\mathbf{k}_v, \mathbf{k}_c) = \frac{\delta_{k_c, k_v + \kappa}}{\Omega} \int_\Omega u^*_{vk_v}(\mathbf{P} + \hbar \mathbf{k}_c)\, u_{ck_c}\, d\mathbf{r}$$

$$+ \sum_{\substack{m \\ (K_m \neq 0)}} \frac{\delta_{k_c, k_v + \kappa + K_m}}{\Omega} \int_\Omega u^*_{vk_v}\, e^{iK_m \cdot \mathbf{r}} (\mathbf{P} + \hbar \mathbf{k}_c)\, u_{ck_c}\, d\mathbf{r}$$

where the contribution from the sum can usually be ignored. Thus, to a good approximation, momentum conservation

$$\mathbf{k}_c = \mathbf{k}_v + \kappa$$

holds. Since, as mentioned in Section 2.2.6, the photon wave vector κ is usually negligible, one finds the 'vertical' transitions of Fig. 2.3 below, and

$$\mathbf{P}_{vc}(\mathbf{k}) \equiv \mathbf{P}_{vc}(\mathbf{k}, \mathbf{k}) \simeq \frac{1}{\Omega} \int_\Omega u^*_{vk} (\mathbf{P} + \hbar \mathbf{k})\, u_{ck}\, d\mathbf{r} \qquad (2.102)$$

Unless the first term in (2.102) is zero, corresponding to a 'forbidden' transition, it is usually much larger than the second term; indeed, if $\mathbf{k}_c$ were strictly equal to $\mathbf{k}_v$, the second term would be zero because of the orthonormality of the Bloch functions for different bands. Although these forbidden transitions have been observed optically, they will not be considered here, so that

$$\mathbf{P}_{vc} \simeq [\mathbf{P}_{vc}(\mathbf{k})]_{\text{allowed}} = \frac{\hbar}{i\Omega} \int_\Omega u^*_{v\mathbf{k}}(\mathbf{r}) \nabla u_{c\mathbf{k}}(\mathbf{r}) \, d\mathbf{r}$$

In order to obtain a quantitative estimate of $\mathbf{P}_{vc}$, one may use the sum-rule for momentum matrix elements between states I and J (Wilson, 1953; Morgan and Landsberg, 1965; Morgan and Galloway, 1967):

$$\frac{1}{m} + \sum_{I(\ne J)} \frac{2|\langle I|P_b|J\rangle|^2}{m^2(E_J - E_I)} = \frac{1}{\hbar^2} \frac{\partial^2 E_J}{\partial k_b^2} = \frac{1}{m_{Jb}}$$

Note that states I and J must refer to the same electron wave vector. Summing over the components $b = x, y, z$, one obtains

$$\sum_{b=x}^{z} \sum_{I(\ne J)} \frac{|\langle I|P_b|J\rangle|^2}{E_J - E_I} = \sum_{b=x}^{z} \frac{m}{2}\left(\frac{m}{m_{Jb}} - 1\right) = \frac{3m}{2}\left(\frac{m}{m_J^*} - 1\right) \qquad (2.103)$$

where

$$\frac{3}{m_J^*} \equiv \sum_{b=x}^{z} \frac{1}{m_{Jb}}.$$

However, the quantity actually required is $|M_{IJ}|^2_{\text{av}}$ as in (2.61). Since

$$(M_{IJ})^2 = |\langle I|\mathbf{e}_{\mathbf{k}\lambda} \cdot \mathbf{P}|J\rangle|^2 = |\langle I| \sum_{b=x}^{z} e_{\mathbf{k}b\lambda} P_b|J\rangle|^2 = |\mathbf{e}_{\mathbf{k}\lambda} \cdot \mathbf{Q}|^2, \qquad (*)$$

where $\mathbf{Q} \equiv \langle I|\mathbf{P}|J\rangle$, one obtains, with the notation of Fig. 2.2,

$$|M_{IJ}|^2_{\text{av}} = \sum_{\lambda=1,2} \int (M_{IJ})^2 \frac{d\Omega}{8\pi} = \int \sum_{\lambda=1,2} |\mathbf{e}_{\mathbf{k}_\lambda} \cdot Q|^2 \frac{d\Omega}{8\pi}$$

$$= \int_0^\pi |\mathbf{Q}|^2 \sin^3\theta \, d\theta \cdot \frac{2\pi}{8\pi}$$

$$= \frac{1}{3} \sum_{b=x}^{z} |\langle I|P_b|J\rangle|^2 \qquad (2.103')$$

Thus (2.103) may be rewritten as a sum rule for radiative matrix elements averaged over polarization:

$$\sum_{\substack{I \\ (\ne J)}} \frac{|M_{IJ}|^2_{\text{av}}}{E_J - E_I} = \frac{m}{2}\left(\frac{m}{m_J^*} - 1\right) \qquad (2.104)$$

(*) The first suffix on $\mathbf{e}$ is the photon wave vector and this should now be denoted by κ, but we have preferred to keep to the form used in (2.58).

Applying this equation for the case when the subscript J refers to a state V in the valence band and neglecting all transitions except those between one conduction band minimum and one valence band maximum, both at the same point in $\mathbf{k}$ space, one obtains

$$| M_{\mathrm{CV}} |^2_{\mathrm{av}} = \frac{m}{2} E_{\mathrm{G}} \left(1 - \frac{m}{m_{\mathrm{V}}} \right) \tag{2.105}$$

where m_{V} is the (negative) isotropic effective mass of the valence band.

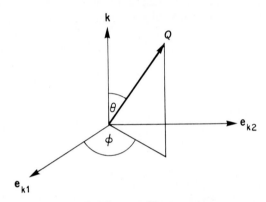

FIG. 2.2. The coordinate system

One may derive an alternative expression (Dumke, 1963) by assuming states in a light hole valence band, a heavy hole valence band and a split-off band, to be labelled by suffices L, H, S. If one uses the wave functions suggested by Kane (1957, 1966) for zinc blende structures this implies that the $\mathbf{k}$-value of the states C, L, H and S lie close to $\mathbf{k} = 0$, and leads to

$$\langle C | P_1 | L \rangle = \langle C | P_2 | H \rangle = \langle C | P_3 | S \rangle, \tag{2.105a}$$

all other components being zero. Hence

$$\langle C | \mathbf{P} | L \rangle = \langle C | \mathbf{P} | H \rangle = \langle C | \mathbf{P} | S \rangle. \tag{2.105b}$$

Now, applying (2.104) with J denoting a state C in the conduction band (assumed to have isotropic effective mass m_{c}), and neglecting other matrix elements,

$$\frac{| M_{\mathrm{CL}} |^2_{\mathrm{av}}}{E_{\mathrm{C}} - E_{\mathrm{L}}} + \frac{| M_{\mathrm{CH}} |^2_{\mathrm{av}}}{E_{\mathrm{C}} - E_{\mathrm{H}}} + \frac{| M_{\mathrm{CS}} |^2_{\mathrm{av}}}{E_{\mathrm{C}} - E_{\mathrm{S}}} = \frac{m}{2} \left(\frac{m}{m_{\mathrm{c}}} - 1 \right).$$

Thus, for states at or near the band extrema,

$$\frac{|M_{\mathrm{CL}}|^2_{\mathrm{av}}}{E_{\mathrm{G}}} + \frac{|M_{\mathrm{CH}}|^2_{\mathrm{av}}}{E_{\mathrm{G}}} + \frac{|M_{\mathrm{CS}}|^2_{\mathrm{av}}}{E_{\mathrm{G}}+\Delta} = \frac{m}{2}\left(\frac{m}{m_{\mathrm{c}}}-1\right)$$

i.e.

$$|M_{\mathrm{CL}}|^2_{\mathrm{av}} = |M_{\mathrm{CH}}|^2_{\mathrm{av}} = |M_{\mathrm{CS}}|^2_{\mathrm{av}} = \frac{mE_{\mathrm{G}}}{6}\left(\frac{m}{m_{\mathrm{c}}}-1\right)\left(\frac{E_{\mathrm{G}}+\Delta}{E_{\mathrm{G}}+\frac{2}{3}\Delta}\right) \quad (2.106)$$

All the quantities in (2.106) are calculated at the **k** value at which (2.105a, b) are valid, and this value is **k** = 0 for GaAs.

This result may also be derived directly from **k.p** perturbation theory (see equation (12) of Kane's 1957 paper). For GaAs, using $m_{\mathrm{c}} = 0.072\, m$, $m_{\mathrm{v}} = 0.5\, m$, $\Delta = 0.33$ eV and $E_{\mathrm{G}} = 1.51$ eV, one finds that expressions (2.105) and (2.106) yield values of $1.5mE_{\mathrm{G}}$ and $2.28mE_{\mathrm{G}}$, respectively; this latter value will be used here.

[ii] *Band-impurity transitions.* In our simple approach, the two transitions labelled as (ii) and (iii) in Fig. 2.1 can be discussed together. Our calculation of these transition probabilities generalizes earlier work in that it can be applied also to excited states of the impurity atom. Previous authors have confined their attention to the ground state (Eagles, 1960; Dumke, 1963; Calloway, 1963; Zeiger, 1964).

For a state $(n, \mathbf{k})$ in a band,

$$\psi_{n\mathbf{k}} = \frac{1}{\sqrt{V}}u_{n\mathbf{k}}(\mathbf{r})\,e^{i\mathbf{k}.\mathbf{r}} = \frac{1}{\sqrt{V}}\sum_{\mathbf{L}} A_n(\mathbf{k}, \mathbf{L})\,e^{i(\mathbf{k}+\mathbf{L}).\mathbf{r}} \quad (2.107)$$

and for the state I of an electron on an impurity energetically nearest to band ν,

$$\psi_I = \sum_{\mathbf{L}}\sum_{\mathbf{k}} A_\nu(\mathbf{k}, \mathbf{L})\, G_{\nu\mathbf{k}}\, e^{i(\mathbf{k}+\mathbf{L}).\mathbf{r}} \quad (2.108)$$

where the **L**'s are lattice vectors in reciprocal space.

The Fourier coefficient $G_{\nu\mathbf{k}}$ is given, for hydrogen-like ground states, by

$$G_{\nu\mathbf{k}} = \frac{8\sqrt{\pi}\,a_\nu^{3/2}}{V(1+a_\nu^2 k^2)^2} \quad (2.109)$$

where a_ν is the quantity appearing in the (normalized) ground state hydrogen-like wave function:

$$\frac{e^{-r/a_\nu}}{(\pi a_\nu^3)^{1/2}}$$

From the usual effective mass theory for impurities one can obtain the impurity ionization energy, E_I, and the value of a_ν:

$$E_I = \frac{\hbar^2}{2m_\nu a_\nu^2}, \quad a_\nu = \frac{\varepsilon\hbar^2}{e^2 m_\nu}, \tag{2.110a, b}$$

where m_ν is the effective mass of band ν. This gives one value of E_I for each m_ν and is often wrong for impurities in GaAs. It is therefore sometimes necessary to treat m_ν and a_ν as unknowns in the above equations, and derive their values from the empirical energy E_I. Some typical values for GaAs are given in Table 2.1.

TABLE 2.1. Typical values of m_ν/m, a_ν and E_I for GaAs.

	1	2	3	4	5	6
	Experimental	Calculated from (2.110) and column 1		Experimental	Calculated from (2.110) and column 4	
	m_ν/m	E_I (eV)	a_ν (cm)	E_I (eV)	m_ν/m	a_ν (cm)
Donors ($\nu = c$)	0·072	0·006	$0·91 \times 10^{-6}$	0·005 (Si)	0·057	$1·15 \times 10^{-6}$
Acceptors ($\nu = v$)	0·5	0·044	$0·13 \times 10^{-6}$	0·04 (Zn)	0·456	$0·14 \times 10^{-6}$

Using the expressions (2.107) and (2.108), the typical band-impurity matrix element is given by

$$
\begin{aligned}
P_{n\mathbf{k},I} &= \int_V \left[\sum_{\mathbf{k}'} \sum_{\mathbf{L}'} A_\nu^* (\mathbf{k}', \mathbf{L}') G_{\nu\mathbf{k}'}^* e^{-i(\mathbf{k}'+\mathbf{L}').\mathbf{r}} \right] \\
&\quad \times \frac{\hbar}{i} \nabla \left[\sum_{\mathbf{L}} \frac{A_n(\mathbf{k}, \mathbf{L})\, e^{i(\mathbf{k}+\mathbf{L}).\mathbf{r}}}{\sqrt{V}} \right] d\mathbf{r} \\
&= \frac{\hbar}{i\sqrt{V}} \int_V \left[\sum_{\mathbf{k}'} \sum_{\mathbf{L}'} A_\nu^* (\mathbf{k}', \mathbf{L}') G_{\nu\mathbf{k}'}^* e^{-i(\mathbf{k}'+\mathbf{L}').\mathbf{r}} \right] \\
&\quad \times \left[\sum_{\mathbf{L}} i(\mathbf{k}+\mathbf{L})\, A_n(\mathbf{k}, \mathbf{L})\, e^{i(\mathbf{k}+\mathbf{L}).\mathbf{r}} \right] d\mathbf{r} \\
&= \sqrt{V}\,\hbar \sum_{\mathbf{L}} (\mathbf{k}+\mathbf{L})\, A_\nu^*(\mathbf{k}, \mathbf{L})\, A_n(\mathbf{k}, \mathbf{L})\, G_{\nu\mathbf{k}}^*
\end{aligned}
$$

where it has been noted that the integral leads to $\mathbf{k} = \mathbf{k}'$ and $\mathbf{L} = \mathbf{L}'$.

Therefore

$$P_{nk,I} = \frac{G_{vk}^*}{\sqrt{V}} \int_V u_{vk}^*(\mathbf{r}) e^{-i\mathbf{k}\cdot\mathbf{r}} \mathbf{P} u_{nk}(\mathbf{r}) e^{i\mathbf{k}\cdot\mathbf{r}} d\mathbf{r}$$

$$= \sqrt{V} G_{vk}^* P_{nk,vk} \tag{2.111}$$

The second factor in (2.111) is a band–band momentum matrix element, as seen by comparing with (2.102), putting $n = v$ and $v = c$. Multiplying (2.111) by a polarization vector and averaging over polarizations, one finds from (2.103′)

$$|M_{nk,I}|_{av}^2 = V|G_{vk}|^2 |M_{nk,vk}|_{av}^2 \tag{2.111'}$$

$$\simeq \frac{64\pi}{V} a_v^3 |M_{nk,vk}|_{av}^2 \tag{2.111''}$$

To evaluate (2.111′), note that in a two-band model the second factor on the right is $|M_{CV}|_{av}^2$ and hence can be estimated from (2.106). Next one needs to determine a_v as illustrated in Table 2.1. One can then estimate the Fourier coefficient (2.109). For the processes of Fig. 2.1 the following identifications have to be made:

Fig. 2.1 (ii): $(nk) \to C$, $(vk) \to V$, $I \to A$

Fig. 2.1 (iii): $(nk) \to V$, $(vk) \to C$, $I \to D$

Note that an order-of-magnitude estimate of eqn. (2.111″) yields

$$|M_{VD}|_{av}^2 \simeq |M_{CA}|_{av}^2 \simeq \frac{10^{-17}}{V} |M_{CV}|_{av}^2$$

However, to make a comparison between the magnitudes of the three different transition probabilities one should multiply in each case (donors and acceptors) by the total number of impurities in the crystal. Hence, denoting the transition probabilities per unit time by T_{IJ} with appropriate values of the subscripts I, J, one obtains

$$T_{VD} \simeq N_D \, 10^{-17} T_{CV}$$

$$T_{CA} \simeq N_A \, 10^{-17} T_{CV}$$

where N_D, N_A are the densities of donors and acceptors, respectively, per unit volume; since these numbers may be of order 10^{16}–10^{19}, the three transition probabilities T_{IJ} are all of roughly similar magnitudes.

Although unimportant in lasers, transitions involving excited hydrogen-like impurity states can be discussed by making the following change:

$$a_v \to a_{v,n'} \equiv n'^2 a_v, \quad E_I \to E_{I,n'} = \frac{n'^2 \hbar^2}{2m_v a_{v,n'}^2}$$

where n' is the principal quantum number involved. Also (2.109) has to be replaced by a more complicated expression (Cohen and Landsberg, 1967).

[iii] *Donor–acceptor transitions.*[3] Using again the wave functions expanded as in (2.108), one obtains the matrix element for this transition (type (iv) of Fig. 2.1) in the following form (allowing for the spatial separation, **R**, of the donor and acceptor):

$$\mathbf{P}_{D,A} = \int_V \left[\sum_L \sum_k A_v^*(\mathbf{k}, \mathbf{L}) G_{vk}^* e^{-i(\mathbf{k}+\mathbf{L}).\mathbf{r}} \right]$$

$$\times \mathbf{P} \left[\sum_{L'} \sum_{k'} A_c(\mathbf{k},'\mathbf{L}') G_{ck'} e^{i(\mathbf{k}'+\mathbf{L}').(\mathbf{r}-\mathbf{R})} \right] d\mathbf{r}$$

$$= \frac{\hbar}{i} \int_V \left[\sum_L \sum_k A_v^*(\mathbf{k}, \mathbf{L}) G_{vk}^* e^{-i(\mathbf{k}+\mathbf{L}).\mathbf{r}} \right]$$

$$\times \left[\sum_{L'} \sum_{k'} i(\mathbf{k}'+\mathbf{L}') A_c(\mathbf{k}', \mathbf{L}') G_{ck'} e^{i(\mathbf{k}'+\mathbf{L}').(\mathbf{r}-\mathbf{R})} \right] d\mathbf{r}$$

$$= \hbar V \sum_L \sum_k (\mathbf{k}+\mathbf{L}) A_v^*(\mathbf{k}, \mathbf{L}) G_{vk}^*$$

$$\times \left[\sum_{L'} A_c(\mathbf{k}+\mathbf{L}-\mathbf{L}', \mathbf{L}') G_{ck+L-L'} e^{-i\mathbf{R}.(\mathbf{k}+\mathbf{L})} \right].$$

As in (2.111), $\mathbf{L} = \mathbf{L}'$ and

$$\mathbf{P}_{D,A} = \hbar V \sum_L \sum_k (\mathbf{k}+\mathbf{L}) A_v^*(\mathbf{k}, \mathbf{L}) A_c(\mathbf{k}, \mathbf{L}) G_{vk}^* G_{ck} \times e^{-i\mathbf{R}.(\mathbf{k}+\mathbf{L})}$$

$$= V \sum_k \mathbf{P}_{vc}(\mathbf{k}) G_{vk}^* G_{ck} e^{-i\mathbf{k}.\mathbf{R}}$$

If one assumes that the main contribution to $\mathbf{P}_{vc}$ comes from the **k** value ($\mathbf{k} = \mathbf{k}_0$ say) at which (2.105a, b) and (2.106) are valid, one obtains:

$$\mathbf{P}_{D,A} \simeq V \mathbf{P}_{vc}(\mathbf{k}_0) \sum_k G_{vk}^* G_{ck} e^{-i\mathbf{k}.\mathbf{R}} \equiv \mathbf{P}_{vc}(\mathbf{k}_0) I \qquad (2.112)$$

The sum I can be evaluated by writing it as the integral of a product of hydrogen ground state wave functions:

$$I = \int_V \frac{\exp\left(-\dfrac{r}{a_v} - \dfrac{|\mathbf{r}-\mathbf{R}|}{a_c} \right)}{\pi(a_c a_v)^{\frac{3}{2}}} d\mathbf{r} \qquad (2.113)$$

---

[3] This subsection is based on Zeiger's paper (Zeiger, 1964).

The evaluation is straightforward but tedious and yields

$$I = \frac{8\alpha^{\frac{3}{2}}}{(\alpha^2-1)^3 \rho} \left\{ e^{-\alpha\rho}[\rho(\alpha^2-1)+4\alpha] + e^{-\rho}[\rho\alpha(\alpha^2-1)-4\alpha] \right\} \quad (2.114)$$

where we have adopted Zeiger's notation of $\alpha \equiv a_v/a_c$, $\rho \equiv R/a_v$.

We have thus obtained the matrix element for the donor–acceptor transition in terms of $\mathbf{R}$, the spatial separation of the impurities. Whenever this matrix element is used it will be needed in the form used in the transition probability, i.e. as some average of the square of the matrix element. $|\mathbf{P}_{D,A}|^2_{av}$ will be evaluated by assuming a random distribution of impurities and integrating over $\mathbf{R}$.

$$|\mathbf{P}_{D,A}|^2_{av} = \int |\mathbf{P}_{D,A}|^2 \frac{d\mathbf{R}}{V} = 4\pi a_v^3 \int_0^\infty |\mathbf{P}_{D,A}|^2 \frac{\rho^2 \, d\rho}{V}$$

This integral is again rather lengthy and the calculation gives:

$$|\mathbf{P}_{D,A}|^2_{av} = |\mathbf{P}_{VC}|^2 \frac{64\pi a_v^3(1+7\alpha+17\alpha^2+7\alpha^3+\alpha^4)}{V(1+\alpha)^7} \quad (2.115)$$

To cast this equation into final form, we observe that

$$|\mathbf{P}_{IJ}|^2 = \sum_{b=x}^{z} |(P_b)_{IJ}|^2 \quad (I,J = D, A \text{ or } V, C)$$

so that by (2.103') the average over polarizations of $|(\mathbf{e} . \mathbf{P})_{IJ}|^2$ is

$$|M_{IJ}|^2_{av} = \tfrac{1}{3}|\mathbf{P}_{IJ}|^2$$

Hence, multiplying (2.115) by $\tfrac{1}{3}$,

$$|M_{D,A}|^2_{av} = |M_{VC}|^2_{av} \frac{64\pi a_v^3(1+7\alpha+17\alpha^2+7\alpha^3+\alpha^4)}{V(1+\alpha)^7} \quad (2.116)$$

A numerical estimate of $|M_{VC}|^2_{av}$ can be obtained from (2.106). The values of α and the a's can be found from (2.110).

2.4.3 Emission rates

The stimulated and spontaneous emission rates are given in (2.92) and (2.93). These equations are summed over all initial and final states (denoted by I and J, respectively) which give rise to a photon of energy E. Converting the resulting sums to integrals, one finds in general that average emission rates per unit volume per unit energy range are given by:

$$r_{st}(E) = \frac{4\mu e^2 E}{Vm^2 \hbar^2 c^3} \int d\mathbf{k}_I \int d\mathbf{k}_J \, g_{IJ}(E)[f_I - f_J]|M_{IJ}(E)|^2_{av} \delta(E_I - E_J - E)$$

$$(2.117)$$

$$r_{sp}(E) = \frac{4\mu e^2 E}{Vm^2 \hbar^2 c^3} \int d\mathbf{k}_I \int d\mathbf{k}_J \, g_{IJ}(E)[f_I(1-f_J)]|M_{IJ}(E)|^2_{av} \delta(E_I - E_J - E)$$

$$(2.118)$$

where f_I and f_J are occupation probabilities, i.e. Fermi–Dirac functions involving the quasi-Fermi levels μ_i and μ_j. $|M_{IJ}(E)|^2_{av}$ is the square of the radiative matrix element for the transition under consideration. The average is over polarizations and directions of the photon wave vector κ, as explained in Section 2.2.6. Because of the lack of exact knowledge of the wave functions entering this factor, it is customary to use some average value assumed independent of initial and final states. It may then be removed from the integral. We now consider, in order, the emission rates associated with the transitions of Fig. 2.1.

The case of band–band transitions has so far received most attention in the literature; even so, there are here two alternative approaches and the full significance of the difference has as yet received little attention. Recall that in the derivation of the matrix element (2.102) we considered the case of conservation of electron wave vector, $\mathbf{k}$. When this matrix element is used in the corresponding emission rates the $\mathbf{k}$-selection rule must be incorporated in the expressions (2.117) and (2.118); this also has the effect of defining the joint density-of-states, as follows. Assuming spherical conduction and valence bands with effective masses m_c and m_v, and band extrema at E_c and E_v, respectively, assumed at the same point in $\mathbf{k}$ space (as in Fig. 2.3), we have that

$$E_I = E_c + \frac{\hbar^2 |\mathbf{k}|^2}{2m_c}$$

$$E_J = E_v - \frac{\hbar^2 |\mathbf{k}|^2}{2m_v}$$

$$E = E_G + \frac{\hbar^2 |\mathbf{k}|^2}{2}\left(\frac{m_c + m_v}{m_c m_v}\right)$$

It follows from the definition of the density-of-states with

$$g_{IJ}(E) = \frac{V}{8\pi^3}$$

that this is given by

$$V\left(\frac{2m_c m_v}{m_c + m_v}\right)^{\frac{3}{2}} \times \frac{(E - E_G)^{\frac{1}{2}}}{2\pi^2 \hbar^3} \tag{2.119}$$

Using this expression, the emission rates for the case of $\mathbf{k}$ selection (Hall,

1963) are given by:

$$r_{st}(E) = A\sqrt{(E-E_G)}\left\{\frac{1}{1+\exp\left[\dfrac{m_v}{m_c+m_v}\left(\dfrac{E-E_G}{kT}\right)-\left(\dfrac{\mu_c-E_c}{kT}\right)\right]}\right.$$

$$\left.-\frac{1}{1+\exp\left[\dfrac{-m_c}{m_c+m_v}\left(\dfrac{E-E_G}{kT}\right)-\left(\dfrac{\mu_v-E_v}{kT}\right)\right]}\right\} \quad (2.120)$$

$$r_{sp}(E) = A\sqrt{(E-E_G)}\left\{\frac{1}{1+\exp\left[\dfrac{m_v}{m_c+m_v}\left(\dfrac{E-E_G}{kT}\right)-\left(\dfrac{\mu_c-E_c}{kT}\right)\right]}\right\}$$

$$\times\left\{1-\frac{1}{1+\exp\left[\dfrac{-m_c}{m_c+m_v}\left(\dfrac{E-E_G}{kT}\right)-\left(\dfrac{\mu_v-E_v}{kT}\right)\right]}\right\} \quad (2.121)$$

where

$$A \equiv \frac{2\mu e^2 E |M_{CV}|^2_{av}}{m^2 \hbar^5 c^3 \pi^2}\left(\frac{2m_c m_v}{m_c+m_v}\right)^{\frac{3}{2}}$$

and μ_c, μ_v are the quasi-Fermi levels of the conduction and valence bands, respectively; A has a typical value of $1\cdot89 \times 10^{39}$ erg^{-1} cm^{-3} sec^{-1} for GaAs. Calculated line shapes for a temperature of 80°K, with $\mu_c - E_c = 5$ meV and $E_v - \mu_v = 11\cdot8$ meV, are shown in Fig. 2.4.

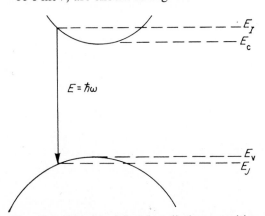

FIG. 2.3. The band–band radiative transition with **k**-selection rule

The assumption of **k** selection implies a reasonably pure semiconductor; however, for a typical injection laser the concentration of impurities can be of the order 10^{18}–10^{19} per cm³. At these high concentrations, impurity scattering will modify the momentum matrix elements involved in interband transitions. The result of such scattering is effectively to relax the **k**-selection rule. Also, the form of the matrix element in this case is more likely to be that of (2.111) for the band-impurity transition. As noted in the discussion following eqn. (2.111′), the transition probabilities and hence the emission rates for these transitions are similar in magnitude to those given by (2.120) and (2.121).

This assumption of no **k**-selection rule was used in a particularly important paper by Lasher and Stern (1964) who also assumed that the impurities had formed an impurity band which, in turn, had merged with the valence band. In view of the lack of more detailed information, it was assumed that the valence-band density-of-states remained unaltered. Assuming also a matrix element independent of initial and final states, the resulting expressions for the emission rates are:

$$r_{st}(E) = C \int_{E_c}^{E_c+E-E_G} (E_I - E_c)^{\frac{1}{2}} (E_v - E_I + E)^{\frac{1}{2}}$$

$$\times \left[\frac{1}{1+\exp\left(\dfrac{E_I - \mu_c}{kT}\right)} - \frac{1}{1+\exp\left(\dfrac{E_I - E - \mu_v}{kT}\right)} \right] dE_I \qquad (2.122)$$

$$r_{sp}(E) = C \int_{E_c}^{E_c+E-E_G} (E_I - E_c)^{\frac{1}{2}} (E_v - E_I + E)^{\frac{1}{2}}$$

$$\times \left[\frac{1}{1+\exp\left(\dfrac{E_I - \mu_c}{kT}\right)} \right] \left[1 - \frac{1}{1+\exp\left(\dfrac{E_I - E - \mu_v}{kT}\right)} \right] dE_I \qquad (2.123)$$

where

$$C \equiv V^2 \frac{8\mu e^2 E(m_c m_v)^{\frac{3}{2}}}{m^2 \hbar^8 c^3 \pi^4} |M_{CV}|^2_{av} |G_v|^2$$

and is equal to $1 \cdot 7 \times 10^{38}$ erg⁻¹ cm⁻³ sec⁻¹, when all energies are expressed in meV, for Zn acceptors in GaAs. In evaluating C, the matrix element used by Lasher and Stern was that given by (2.111) except for the fact that, for greater simplicity, that part of the Fourier coefficient (2.109) which involves k was ignored. The resulting line shapes for this case of no **k**-selection rule are given in Fig. 2.5 for the same parameters as Fig. 2.4; the two figures

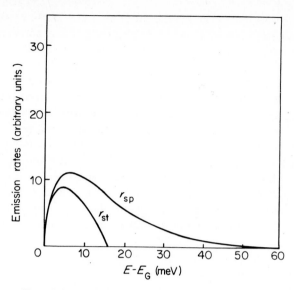

FIG. 2.4. Emission rates with **k**-selection rule calculated from eqns. (2.120) and (2.121)

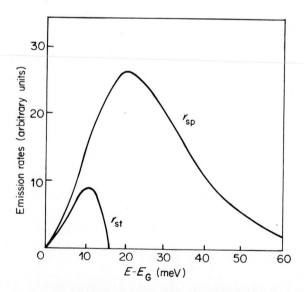

FIG. 2.5. Emission rates without **k**-selection rule calculated from eqns. (2.122) and (2.123)

are drawn to the same scale for the purpose of comparison. Methods for approximating the integrals in (2.122) and (2.123) and obtaining explicit expressions for $r_{st}(E)$ and $r_{sp}(E)$ have been described by Marinelli (1965) and Unger (1967a).

Clearly the two most immediate points of uncertainty in the above theory involve: (*a*) the matrix element whose exact form is as yet unknown in the presence of a high concentration of impurities, and (*b*) the density-of-states function for a high concentration of impurities; this latter problem, however, has received a great deal of attention in recent years.[4] The simple assumption that the band density-of-states remains unaltered (mentioned above) is rather dubious, as the usual effect of a high density of impurity states near an energy band is to form a band tail whose density-of-states may be exponential, Gaussian or of some more complicated form. The simplest density-of-states for such a tail is thus:

$$g_c = g \exp\left(\frac{E_I - E_c}{E_0}\right) \tag{2.124}$$

where g is a constant and the suffix 'c' here indicates that we are considering a conduction band tail. Pankove (1965) has given experimentally deduced values for the parameter E_0 from optical absorption measurements. Lasher and Stern also calculated a stimulated line shape using this density-of-states with the value $E_0 = 15$ meV and the result is shown in Fig. 2.6 together with the line calculated using a parabolic density-of-states and the experimental results of Nelson and coworkers (1963) and Burns and Nathan (1964).

In a more recent paper, Stern (1966b) used the Gaussian band tail result obtained by Kane (1963a) where, for example, the conduction band tail density-of-states is

$$g_c = \frac{m_c^{\frac{3}{2}}}{\pi^2 \hbar^3} (2\eta_c)^{\frac{1}{2}} y\left[\frac{E_I - E_c}{\eta_c}\right] \tag{2.125}$$

with

$$y(x) = \frac{1}{\sqrt{\pi}} \int_{-\infty}^{x} (x-z)^{\frac{1}{2}} e^{-z^2} dz \tag{2.126}$$

and

$$\eta_c = \frac{e^2}{\varepsilon} (4\pi N_D L)^{\frac{1}{2}} \tag{2.127}$$

---

[4] See, for example, Parmenter, 1956; Lax, 1952; Edwards, 1961; Klauder, 1961; Lax and Phillips, 1958; Frisch and Lloyd, 1960; Wolff, 1962; Kane, 1963a, b; Halperin and Lax, 1966a, b; Jones and Lukes, 1969.

Here ε is the static dielectric constant, N_D is the concentration of donors and L is the screening length which must be calculated from the positions of the quasi-Fermi levels in the non-equilibrium situation under consideration:

$$\frac{1}{L^2} = \frac{4\pi e^2}{\varepsilon} \left(\frac{dn}{d\mu_c} - \frac{dp}{d\mu_v} \right)$$

Some spontaneous line shapes calculated by Stern using this expression are shown in Fig. 2.7.

Once again, Unger (1967b) has derived analytic approximations for these curves. Other calculations have been made by Lucovsky (1965, 1966) using a simple Gaussian density-of-states for the band tails with Kane's expression (2.127) for the band tail spreading energy η_c; Lucovsky was able to obtain a reasonable fit to experimental results using this simple approximation. A mechanism also exists for the formation of low-energy tails (such as those appearing in Fig. 2.7), for low impurity concentrations (Landsberg, 1966).

To complete the discussion, we demonstrate an approximate method suggested by Zeiger (1964) for comparing the emission rates due to donor–acceptor transitions with those due to donor–valence band transitions (possibly the two most likely transitions involved in the injection laser).

The stimulated emission rate for the donor–acceptor transition is, from (2.116),

$$r_{st}^{d-a}(E) = \frac{4\mu e^2 E}{m^2 \hbar^2 c^3} |M_{CV}|_{av}^2 \frac{64\pi a_v^3 (1 + 7\alpha + 17\alpha^2 + 7\alpha^3 + \alpha^4)}{(1+\alpha)^7} N_A N_D$$

$$\times \iint g_D(E_D) g_A(E_A) [f_D(E_D) - f_A(E_A)] \, dE_D \, dE_A \, \delta(E_A - E_D - E)$$

$$(2.128)$$

where: (a) N_D, N_A are total numbers of donors and acceptors, (b) f_D, f_A are occupation probabilities for donors and acceptors, (c) g_D, g_A are shape functions for the energy distribution of donors and acceptors, which are normalized so that $\int g_D(E_D) \, dE_D = \int g_A(E_A) \, dE_A = 1$ and (d) E_D, E_A are energy levels of donor and acceptor, respectively.

The stimulated rate for the donor–valence band transition is

$$r_{st}^{d-v}(E) = \frac{4\mu e^2 E}{V m^2 \hbar^2 c^3} |M_{CV}|_{av}^2 \frac{V}{2\pi^2} \left(\frac{2m_v}{\hbar^2} \right)^{\frac{3}{2}} \iint (E_v - E_J)^{\frac{1}{2}} g_D(E_D) N_D$$

$$\times \frac{2^{\frac{3}{2}} \pi \hbar^3 [f_D(E_D) - f_v(E_J)] \, dE_J \, dE_D}{[m_c(E_c - E_D)]^{\frac{3}{2}} \left[1 + \frac{m_v(E_v - E_J)}{m_c(E_c - E_D)} \right]^4} \times \delta(E_J - E_D - E)$$

$$(2.129)$$

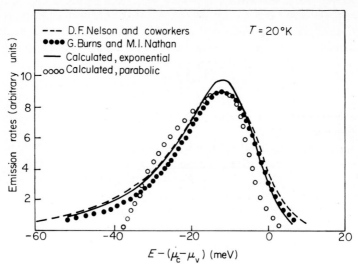

FIG. 2.6. Experimental and theoretical emission rates from
GaAs, after Lasher and Stern (1964)

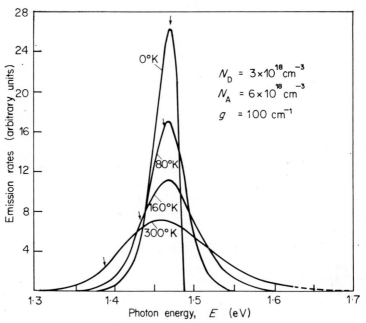

FIG. 2.7. Theoretical spontaneous emission rates, after Stern
(1966b)

Zeiger's simplifying assumptions at this point are: (a) that f_A and f_V (within energy $|(E_c - E_D) m_c / m_v|$ of the top of the valence band) are roughly constant, (b) that the donor energy distribution is much broader than that of the acceptors, so that $g_A(E_A)$ is roughly a δ function and (c) that $\dfrac{m_c}{m_v} \ll 1$, so that

$$\frac{(E_v - E_J)^{\frac{1}{2}}}{\left[1 + \dfrac{m_v(E_v - E_J)}{m_c(E_c - E_D)}\right]^4}$$

is roughly a δ function. The integrals in (2.128) and (2.129) can then be performed and we obtain:

$$r_{st}^{d-a}(E) = \frac{4\mu e^2 E}{m^2 \hbar^2 c^3} |M_{CV}|_{av}^2 \frac{64\pi a^3(1 + 7\alpha + 17\alpha^2 + 7\alpha^3 + \alpha^4)}{(1+\alpha)^7} N_A N_D$$
$$\times g_D(E_D)[f_D(E_D) - f_A] \tag{2.130}$$

$$r_{st}^{d-v}(E) = \frac{4\mu e^2 E}{m^2 \hbar^2 c^3} |M_{CV}|_{av}^2 2N_D g_D(E_D)[f_D(E_D) - f_V] \tag{2.131}$$

From these, if one assumes f_V is of the same order as f_A, it is seen that $r_{st}^{d-a}(E)$ will exceed $r_{st}^{d-v}(E)$ when

$$N_A > \frac{(1+\alpha)^7}{32\pi a^3(1 + 7\alpha + 17\alpha^2 + 7\alpha^3 + \alpha^4)} \tag{2.132}$$

Using the values $E_C - E_D = 0.006$ eV, $E_A - E_V = 0.04$ eV, with other parameters appropriate to GaAs, one finds that the donor–acceptor transition rate is greater than the donor–valence band rate for concentrations of acceptors in excess of about 4×10^{18} cm^{-3}. In fact the result (2.132) was not obtained specifically in this form by Zeiger but has since been pointed out by Lax (1964) who also notes that a non-random pairing of donors and acceptors would tend to enhance this form of transition. However, the concentration of impurities mentioned above is probably sufficient to form an impurity band tail on the valence band and hence the distinction between the two types of transitions discussed is not important.

2.5 Threshold relations

In Sections 2.2, 2.3 and 2.4, we have dealt with the various physical processes responsible for emitting and absorbing radiation in a semiconductor and with the quantum-mechanical and statistical descriptions of these processes. We now come to the rather more specialized problems associated with the semiconductor laser and the radiation which it emits.

The conventional theory of solid-state and gaseous lasers[5] is by now well established and we therefore deal only with aspects which are specific to the semiconductor laser. Thus no detailed discussion is given here of the kinetic or master equations for the time development of a laser system, of Fabry–Perot interferometer theory, coherence problems, etc., since these are common to all laser systems and covered adequately in other books.

By its definition the laser is a device which amplifies light by stimulated emission of radiation; it therefore demands the existence of a suitable amount of stimulated emission and the provision of some form of feedback mechanism to ensure that the radiation passes through the device many times and undergoes amplification. In the injection laser the first of these requirements is achieved by the use of a forward-biased p–n junction and the second by the construction of a resonant structure in the form of two parallel reflecting surfaces perpendicular to the plane of the junction; this part of the device is known as a Fabry–Perot interferometer.

2.5.1   The p–n junction

In Section 2.3 it was shown that the condition for negative absorption (or, equivalently, for population inversion) was given by (2.95) and (2.98):

$$\mu_i - \mu_j > h\nu \tag{2.133}$$

where μ_i, μ_j are quasi-Fermi levels associated with the upper and lower groups of states, respectively, and ν is the frequency of the photon involved. Since $h\nu$ is of the order of the energy gap, (2.133) implies for band–band transitions that at least one band is degenerate. In fact, a simple means of achieving the condition (2.133) is at the junction of a degenerate n-type semiconductor with a degenerate p-type specimen of the same material. The energy diagram of such a junction biased in the forward direction is shown in Fig. 2.8.

The usual procedure for producing p–n junctions of the form used in GaAs lasers are diffusion, epitaxy and alloying. As a general rule, diffusion tends to give linearly graded junctions whilst the remaining two methods are more suitable for abrupt junctions. The impurity gradient in a given junction may be calculated from capacitance–voltage measurements (Pilkuhn, 1968). A typical diffused junction may have a concentration gradient of the order 10^{23} cm^{-4}. Clearly, such features as the method of fabrication and impurity concentration have important effects upon laser parameters such as the active region width, and upon the carrier transport

---

[5] For general reviews of the subject see the following: Birnbaum, 1964; Lengyel, 1966; Pátek, 1967; Yariv and Gordon, 1963; Mashkevich, 1967.

mechanisms which to a certain extent determine the injection efficiencies and recombination paths.

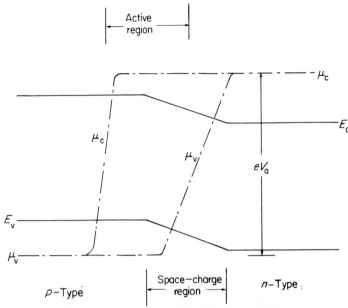

FIG. 2.8. Idealized model of the *p–n* junction

Stern (1966a) has given a simple method of estimating the thickness of the active region as follows. The required dimension may be assumed to be of the same order as the diffusion length for electrons, L_n, defined as

$$L_n = (D_n \tau_n)^{\frac{1}{2}} \qquad (2.134)$$

where D_n is the diffusion constant for a non-degenerate electron gas, and τ_n the steady-state lifetime, given as

$$\tau_n \simeq \frac{1}{B\bar{p}} \qquad (2.135)$$

Here $\bar{p}$ is the average hole concentration and B is a constant associated with the recombination. It is further assumed that $\bar{p}$ is given by

$$\bar{p} \simeq GL_n \qquad (2.136)$$

where G is the impurity gradient at the junction. Combining eqns. (2.134), (2.135) and (2.136) we obtain

$$L_n \simeq \left(\frac{D_n}{GB}\right)^{\frac{1}{3}} \qquad (2.137)$$

Using typical values for GaAs of $G = 4 \times 10^{22}$ cm^{-4}, $B(77^\circ\text{K}) = 4 \times 10^{-10}$ cm^3/sec (Dumke, 1963) and $D_n = 15$ cm^2/sec, Stern finds that $L_n \simeq 1\mu$ which agrees well with values of the active region width deduced from the electromagnetic theory of the GaAs junction laser (Stern, 1965). Note that the diffusion coefficient which has been used here is that appropriate to electrons. This is much larger than D_p, the diffusion coefficient for holes, because the electron effective mass is much smaller than that of the holes in GaAs. Hence the junction has also a basic asymmetry which will be noted later in this subsection.

Broadly speaking, the conduction mechanisms available at the junction are those associated with thermal diffusion and tunnelling (see Fig. 2.9). In the case of the former, there are two possibilities: recombination in the space-charge region, or diffusion of electrons right across the space-charge region and subsequent recombination in the p-type region. In the case of tunnelling there are also two alternatives: the electron may tunnel horizontally from the n region to an impurity state in the p region (with or without thermal equilibration) i.e. band-tail filling, with subsequent recombination, or there may be diagonal tunnelling consisting of a single energy-conserving process in which no intermediate states are needed even though the initial and final states are spatially separated. Also, in both cases, a certain amount of tunnelling by holes in the opposite direction is to be expected. It should be noted, however, that *near threshold* tunnelling is a relatively unimportant mechanism. In this region the photon energy is approximately that of the band gap and hence the electric field at the active region is low; for this reason there was no need to introduce the electric field into the matrix elements calculated in Section 2.4. In general, *below threshold*, the dominant transport mechanism is determined by the impurity gradient, doping and injection levels and temperature of the junction; the interpretation of laser performance in terms of any one specific mechanism is frequently very doubtful. An analysis of the situations which may arise has been given by Lamorte, Caplan and Gonda (1967).

The presence of these different transport mechanisms together with the phenomena of impurity bands and band tails in heavily-doped junctions lead to varying current–voltage characteristics (the applied voltage, V_a, is equal to the separation of the quasi-Fermi levels, μ_c and μ_v, in Fig. 2.8). The problem of determining the I–V characteristics for the various cases of low and high injection levels, light and heavy doping, etc., is well known (Sah, Noyce and Shockley, 1957; Evans and Landsberg, 1963; Morgan, 1965); however, much of the conventional theory is not applicable to the (usually degenerate) junction used in the injection laser. In general, when

(a)

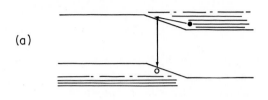

(b)

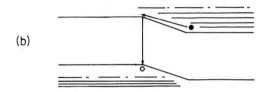

(c)

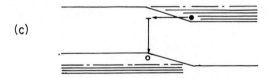

(d)

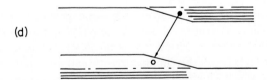

FIG. 2.9. The conduction mechanisms: (a) Diffusion with recombination in space-charge region. (b) Diffusion with recombination on *p*-side. (c) Horizontal tunnelling-assisted recombination. (d) Diagonal tunnelling-assisted recombination

diffusion is the dominant conduction mechanism, the non-degenerate I–V relation has the form

$$I = I_0\left[\exp\left(\frac{eV_a}{\beta kT}\right) - 1\right] \tag{2.138}$$

For recombination in the p region, β usually has the value unity, whilst for recombination in the space-charge region, $\beta < 2$. Morgan (1965) has calculated the critical value of the bias, V_a, at which β changes from one to two, and also another critical bias above which β becomes larger than two; these critical values are also found experimentally.

When tunnelling dominates, the *non-degenerate* I–V relation is also exponential (Leite and coworkers, 1965):

$$I = I_0\exp\left(\frac{eV_a}{\phi}\right) \tag{2.139}$$

with the significant difference that ϕ is independent of temperature. However, a similar relation to (2.139) is also found theoretically for degenerate statistics and band-filling for heavily doped material when exponential band tails with density-of-states given by (2.124) are assumed. In this case one obtains (Unger, unpublished; Adams, 1969):

$$I = I_0\exp\left(\frac{eV_a}{E_0}\right)$$

where E_0 is a parameter determined by the screening length and varying very little with temperature. It should be noted that both (2.138) and (2.139) are only applicable below lasing threshold and without the phenomenon of stimulated emission.

A further aspect of the junction problem is the degree of compensation present on the p side. Winogradoff and Kessler (1964, 1966, 1967) have observed that there exists a critical optimum degree of compensation, resulting in a reduction of threshold currents and an increase in the radiation power outputs. They have also carried out a theoretical investigation of this effect using, however, only non-degenerate statistics and parabolic bands. Briefly their arguments are: (*a*) that the density-of-states in the conduction band of normal uncompensated GaAs is too high for the production of an inverted population at practical current values, and (*b*) that there is a critical separation of the quasi-Fermi levels in the junction leading to a maximum recombination rate. The presence of compensation has the effect of forming a low density-of-states tail to the conduction band with the results: (*a*) of facilitating the formation of an inverted population, and (*b*) of reducing the band gap and junction width (assuming the slope of the potential barrier remains constant).

A complete solution of the junction problem for the injection laser, allowing for degeneracy, injection mechanisms, band tails, etc., is still outstanding.

2.5.2   *The condition for oscillation in a Fabry–Perot interferometer*

It has been mentioned above that the simplest form of resonant cavity to be used for the amplification of electromagnetic radiation consists of two parallel reflecting surfaces. For the injection laser this is achieved by cleaving one pair of faces of the diode (usually along the (110) plane for GaAs); the semiconductor–air surface provides all the reflectivity necessary, due to the high refractive index (μ) of the material ($\simeq 3.6$ for GaAs). Usually the other faces of the diode are roughly finished to discourage the formation of electromagnetic modes propagating in undesired directions. However, it is possible to produce diodes with all four faces cleaved; these, as well as other shapes of diode, have been used in attempts to produce devices with lower threshold currents.

The detailed theory of laser oscillations in Fabry–Perot cavities for various sizes and shapes of the reflectors and for the cases when these are not perfectly parallel nor perfectly plane, has been extensively studied (Schawlow and Townes, 1958; Fox and Li, 1961; Boyd and Gordon, 1961; Kotik and Newstein, 1961); we shall confine our discussion to the idealized case of two perfectly plane, parallel, infinite-width mirrors, with real reflectivities R_1, R_2, separated by a distance L, as indicated in Fig. 2.10.

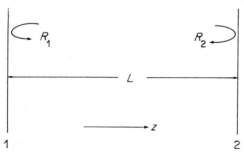

FIG. 2.10. Schematic diagram of the Fabry–
Perot cavity

Consider the field to be given by a plane electromagnetic wave, of complex wave vector **q**, propagating in the z direction; the energy carried by the wave is then given by the Poynting vector and is proportional to

e^{2iqz}, since **E** and **H** are both assumed to vary as e^{iqz}. We now require the condition for this wave to be amplified by successive journeys through the cavity and reflections at each mirror; thus after one complete round trip we must at least require the wave to have the same form: i.e.

$$R_1 e^{2iqL} R_2 e^{2iqL} = 1 \qquad (2.140)$$

From Maxwell's equations, if the electric and magnetic fields vary as $\exp(i\omega t)$, one finds:

$$q = \frac{\omega}{c}\sqrt{\varepsilon\mu_0} \equiv \frac{\omega}{c}(\mu + i\kappa) \qquad (2.141)$$

where ε is the complex dielectric constant, μ_0 is the magnetic permeability and μ is the (real) refractive index; κ is a real quantity which governs absorption and is sometimes termed the extinction coefficient. Taking the square root of both sides of eqn. (2.140) and using (2.141), one obtains:

$$e^{2iqL} = \exp\left[\frac{2iL\omega\mu}{c} - \frac{2\omega\kappa L}{c}\right] = \frac{1}{\sqrt{(R_1 R_2)}} \equiv \frac{1}{R},$$

say. Taking the phase and amplitude parts of this equation, respectively,

$$\frac{2L\omega\mu}{c} = 2M\pi, \qquad M \text{ some integer}$$

i.e.

$$L = \frac{Mc}{2\nu\mu} \equiv \frac{M\lambda}{2\mu}, \qquad (2.142)$$

and

$$\frac{2\omega\kappa}{c} = -\frac{1}{L}\ln\frac{1}{R}. \qquad (2.143)$$

Equation (2.142) expresses a quantization condition of the familiar form

$$\frac{L}{\lambda/2\mu} = \text{an integer},$$

where λ is throughout the vacuum wavelength of the wave in the device. An electromagnetic mode in the cavity will be specified by a value of M together with a direction of polarization perpendicular to the z axis. In the sequel, however, the above equations will be combined with results for absorption and emission (See Subsection 2.3.3, pp. 24–25) These latter results imply that an average over directions of polarization has been performed. Consequently modes need thereafter be specified only by the integer M.

To discuss (2.143) it is convenient to denote by α' a loss term per unit length (for which no single analytic expression is available). It may contain contributions from reabsorption and free-carrier absorption in the 'lossy' p and n regions, i.e. diffraction losses and possibly absorption effects in the active region as a result of either free carriers or incomplete population inversion. The quantity on the left of (2.143) is an effective absorption coefficient (Moss, 1959)—a negative quantity in the active region for all cases of interest. However, since κ is the exponent governing the decay of a plane wave, it includes also the effects of the losses α'. The relation with the absorption coefficient defined in Section 2.3.4 is therefore

$$\alpha(E)+\alpha'(E) = \frac{2\omega\kappa}{c} = -\frac{1}{L}\ln\frac{1}{R}$$

where the suffices i,j indicating the two groups of quantum states which contribute to the transitions, have been suppressed.

These results are for plane waves of any allowed photon energy E, and will now be applied to the energy E_m at which the stimulated emission is a maximum. The energy E_m can refer to any of the modes M given by (2.142) which lie near this maximum, since lasing may be expected to occur near this energy value. Thus one obtains

$$-\alpha(E_m) = \alpha'(E_m)+\frac{1}{L}\ln\frac{1}{R} \tag{2.144}$$

or, using (2.94),

$$r_{\text{st}}(E_m) = \frac{-\mu^2 E_m^2}{\pi^2 c^2 \hbar^3}\alpha(E_m) = \frac{\mu^2 E_m^2}{\pi^2 c^2 \hbar^3}\left[\alpha'(E_m)+\frac{1}{L}\ln\frac{1}{R}\right] \tag{2.145}$$

It is easily seen that these equations represent a threshold for lasing, for if the '=' in (2.144) or (2.145) were '<', the wave will not suffer sufficient amplification to overcome the losses. The right-hand side of (2.144) is the upper limit ('saturation value') of the 'gain' $-\alpha(E_m)$.

Thus two conditions, (2.133) and (2.144), have been deduced for lasing. As the excitation of the device is increased, there are three distinct operating regions separated by these two conditions. First, at low injection level, spontaneous emission occurs; next, as the excitation is increased, (2.133) will be satisfied and stimulated emission commences; finally, when sufficient excitation is applied, condition (2.144) will be satisfied and a lasing threshold may be said to have been achieved.

2.5.3 A simple expression for the threshold current density

One of the main problems associated with the injection laser is that of predicting the minimum current density through the device for which the threshold condition (2.144) is satisfied. A simple expression for such a threshold current density can be derived as follows. The kinetic equations governing the rate of change of the electron density, n, in the active region and the number of quanta per electromagnetic mode, N_M, may be written (Vilms and coworkers, 1966; Vinetskii and coworkers, 1965) as:

$$\frac{dn}{dt} = R_{\text{inj}} - \frac{R_{\text{sp}}}{\eta_i} - \frac{1}{WLd} \sum_{\substack{M \\ (E_M \sim E_m)}} \left[\frac{r_{\text{st}}(E)}{\phi(E)}\right]_{E_M} N_M \qquad (2.146)$$

$$\frac{dN_M}{dt} = -\frac{N_M}{\tau_M} + WLdR_{\text{sp},M} + \left[\frac{r_{\text{st}}(E)}{\phi(E)}\right]_{E_M} N_M \qquad (2.147)$$

where the following notation has been used:

$R_{\text{inj}} = \dfrac{I}{eWLd}$ = injection rate of electrons per second per unit volume,

η_i = quantum efficiency of radiative recombination,

W = width of device,

L = length of device (between reflectors),

d = penetration depth of the electromagnetic wave away from the junction,

$\phi(E)$ = number of modes per unit volume per unit energy,

        = $2\mathcal{N}/V$ where $\mathcal{N}$ is defined in eqn. (2.60),

τ_M = photon lifetime in a lasing mode,

$$\text{i.e.} \frac{1}{\tau_M} = \frac{c}{\mu}\left[\alpha'(E_m) + \frac{1}{L}\ln\frac{1}{R}\right]$$

E_m = energy corresponding to the peak value of the stimulated emission,

$r_{\text{st}}, R_{\text{sp}}$ are as defined in (2.90) and (2.91),

$R_{\text{sp},M}$ is the total spontaneous emission rate into mode M.

The first term on the right-hand side of (2.146) represents additions to the electron concentration from the injection current; the second and third terms, respectively, represent losses due to spontaneous and stimulated radiative recombination. Non-radiative transitions represent an additional loss in (2.146) which is, however, included only in an approximate way through the quantum efficiency. Similarly, the second and third terms on the right of eqn. (2.147) account for additions to the mode population due to spontaneous and stimulated emission, whilst the first term represents the losses from the mode. For completeness, we note that the loss term may be more conventionally expressed in terms of the *laser quality factor*, Q_M, defined as

$$Q_M \equiv \omega_M \tau_M \equiv \frac{E_M \tau_M}{\hbar}$$

Clearly, the larger the Q the smaller the loss term in equation (2.147).

In this article only steady-state situations will be considered. In that case (2.147) yields

$$N_M = \frac{WLd\,R_{\text{sp},M}}{\dfrac{1}{\tau_M} - \dfrac{r_{\text{st}}(E_M)}{\phi(E_M)}} = \frac{WLd\,R_{\text{sp},M}}{\dfrac{c}{\mu}\left[\alpha(E_M) + \alpha'(E_M) + \dfrac{1}{L}\ln\dfrac{1}{R}\right]}$$

It is clear that when condition (2.144) for lasing is satisfied for mode M there will be a strong maximum in the corresponding occupation number N_M. This gives rise to the familiar 'lasing peaks'. The frequency at which this occurs will be termed the *laser frequency* for that mode.

The steady-state solution of (2.146) yields

$$R_{\text{inj}} = \frac{R_{\text{sp}}}{\eta_i} + \frac{1}{WLd}\sum_M \left[\frac{r_{\text{st}}(E)}{\phi(E)}\right]_{E_M} N_M \tag{2.148}$$

Prior to the onset of lasing, both r_{st} and the mode population N will be very small and the second term on the right in eqn. (2.148) can usually be ignored;[6] hence, near the threshold, it is customary to use the approximate expression for the current density:

$$j = \frac{I}{WL} = \frac{edR_{\text{sp}}}{\eta_i} \tag{2.149}$$

If the spontaneous emission is characterized by an empirical or theoretical linewidth ΔE at half-maximum (cf. the various line shapes illustrated in Section 2.4), it is convenient to define:

$$\gamma = \frac{R_{\text{sp}}}{\Delta E \times r_{\text{st}}(E_m)} \tag{2.150}$$

If the useful (stimulated) radiation is maximized,

$$R_{\text{sp}} \sim \int r_{\text{st}}(E)\,dE \sim r_{\text{st}}(E_m)\,\Delta E,$$

---

[6] Note, however, that Vinetskii and coworkers have considered the solutions of equations similar to (2.146) and (2.147). Retaining all the terms in (2.146) and (2.147), they determined those lasing modes in which the number of photons is maximal ('singular' modes).

3

i.e. $\gamma \sim 1$. This is possible at absolute zero temperature. In other cases γ will exceed unity and it is thus a dimensionless 'demerit' factor introduced to account for line shape and temperature effects; some expressions for it will be obtained below. Note that this factor is not included by most authors and is claimed (Lamorte and coworkers, 1967) to be of order unity for GaAs in the temperature range from 77 to 300°K.

Using (2.145), (2.149) and (2.150), one obtains (Lasher, 1963; Yariv and Leite, 1963, 1964; Lasher and Stern, 1964):

$$j_t = \frac{ed}{\eta_i} \frac{\gamma \Delta E \mu^2 E_m^2}{\pi^2 c^2 \hbar^3} \left(\alpha' + \frac{1}{L} \ln \frac{1}{R} \right)$$

$$= \frac{8\pi ed\mu^2 \Delta v \gamma}{\eta_i \lambda_m^2} \left(\alpha' + \frac{1}{L} \ln \frac{1}{R} \right) \qquad (2.151)$$

where $\lambda_m = c/v_m$ = wavelength in vacuo corresponding to the peak value of the stimulated emission.

The use of eqn. (2.151) to predict threshold currents is complicated by lack of knowledge as to the exact values of the parameters involved. Writing

$$\alpha' = \alpha_0 + \alpha_{\text{diff}},$$

where α_0 is the absorption coefficient in the active region due to free carrier effects, etc., and α_{diff} is the absorption coefficient due to diffraction losses, then the value of this latter parameter, together with the value of d, may be estimated from the electromagnetic theory of the laser (to be dealt with in Section 2.6). However, the values of α_0, η_i and ΔE to be used are still uncertain. Lasher (1963) assumes values of $d = 1 \cdot 1 \times 10^{-3}$ cm, $\mu = 4$, $\gamma = \eta_i = 1$, $E = 1 \cdot 47$ eV, $\Delta E = 0 \cdot 025$ eV, $R = 30\%$, $L = 1$ mm, $\alpha_{\text{diff}} = 7$ cm^{-1}. Neglecting α_0, he arrives at a value of $j_t = 830$ A/cm^2 which compares well with an early experimental result of 700 A/cm^2 obtained at a temperature of $4 \cdot 2$°K (Quist and coworkers, 1962). Note that Lasher's value for d has to be large to account for experiment because he has taken γ to be one; as we shall see below, a larger value of γ is frequently more appropriate even at low temperatures.

One could rewrite eqn. (2.151) using (2.144) in the form

$$j_t = -\frac{\alpha(E_m)}{\beta}.$$

This shows how β may be found from experiment whereas (2.151) gives the theoretical expression.

However, another disadvantage of eqn. (2.151) is the lack of any explicit temperature dependence. This aspect of the threshold current is of the utmost importance, since it is found experimentally that the threshold increases with temperature at such a rate that at high temperatures continuous working is not possible, and pulsed operation must be employed.

2.5.4 The 'demerit' factor γ

It is of interest to examine the 'demerit' factor γ introduced in equation (2.150):

$$\gamma(T) \equiv \frac{\int r_{sp}(E)\, dE}{\Delta E r_{st}(E_m)}$$

$$= \frac{\int r_{sp}(E)\, dE}{\Delta E r_{sp}(E_m)} \times \left[\frac{1}{1 - \exp\left(F_j - F_i + \dfrac{E_m}{kT}\right)} \right] \quad (2.152)$$

where Eqn. (2.93) has been used. For a realistic case the last factor in (2.152) can be (Stern, 1966b) about 1·9, corresponding to $T = 300°$K. Assuming this, one has

$$\gamma(300) = 1\cdot9\gamma(0) \quad (2.153)$$

For a simple Lorentzian line shape

$$r_{sp}^{(L)}(E) \equiv \frac{\Delta E}{2\pi} \times \frac{D}{\Delta E^2/4 + (E - E')^2},$$

where D is an energy-independent constant, one finds

$$\gamma^{(L)}(300) = 1\cdot9 \times \frac{\pi}{2} = 2\cdot98. \quad (2.154)$$

Similarly, for a Gaussian line shape:

$$r_{sp}^{(G)}(E) \equiv \frac{(\ln 2)^{\frac{1}{2}}}{\pi^{\frac{1}{2}}} \times \frac{2D}{\Delta E} \times \exp\left\{ -\left[\left(\frac{E - E'}{\Delta E}\right) 2(\ln 2)^{\frac{1}{2}} \right]^2 \right\},$$

so that

$$\gamma^{(G)}(300) = \frac{1\cdot9\pi^{\frac{1}{2}}}{2(\ln 2)^{\frac{1}{2}}} = 2\cdot02. \quad (2.155)$$

Using his calculated line shapes, together with appropriate parameters for GaAs lasers at 300°K, Stern (1966b) finds (2.153) valid. In his case,

$$\gamma^{(S)}(300) = 1\cdot9 \times 5\cdot11 = 9\cdot7 \quad (2.156)$$

Hence, in all cases much of the excitation is 'wasted' in producing spontaneous emission.

2.5.5   Temperature dependence of the threshold current

Using Lasher's result (2.151), the first attempt to incorporate an explicit temperature dependence was made by Moll and Gibbons (1963), who obtained an expression for α_0 on the basis of a simple model involving recombination through an acceptor level and using non-degenerate statistics. Their calculations showed reasonably good agreement with early experimental results at low temperatures but yielded a theoretical temperature dependence $j_t \sim T^{\frac{3}{2}}$, which now appears rather unlikely in view of the T^3-law which is usually observed near 300° K. This is also the disadvantage of a similar result by Mayburg (1963). This paper deals with the condition for population inversion (2.133) and uses non-degenerate statistics assuming $(E_c - \mu_c) \simeq 2kT \simeq (\mu_v - E_v)$. The threshold current densities are then calculated for the cases: (*a*) of recombination in the space-charge region, and (*b*) of recombination on the *n* side of the junction. Both cases yield a 3/2-power law for the temperature dependence.

A rather more successful theory of threshold currents is that arising from the band-filling model of Nelson and coworkers (1963). According to this model, recombination takes place on the *p* side of the junction where the donors have formed an exponential band tail. The injection mechanism for electrons from the *n* region consists then of tunnelling at constant energy through impurity states. If we now assume that at $T = 0°K$, the expression (2.151) for j_t is correct (with $\gamma = 1$), then at this temperature all states up to the energy μ_c are occupied by electrons. As the temperature increases, thermal broadening of the electron population occurs and an additional current density, j_1, is required to fill all the states up to μ_c. Then, assuming this additional current density is that needed to fill all states between μ_c and $\mu_c + kT$, and that the band tail density-of-states is given by (2.124),

$$j_1 \simeq \frac{ed}{\tau'} \int_{\mu_c}^{\mu_c + kT} g \exp\left(\frac{E_I - E_c}{E_0}\right) dE_I$$

$$= \frac{edg E_0}{\tau'} \left[\exp\left(\frac{\mu_c + kT - E_c}{E_0}\right) - \exp\left(\frac{\mu_c - E_c}{E_0}\right)\right]$$

where τ' is the radiative lifetime.

Taking $\mu_c \simeq E_c$ in accordance with condition (2.133), one obtains for the total current density at threshold (Dousmanis and coworkers, 1964):

$$j_t = \frac{8\pi e d \mu^2 \Delta \nu}{\eta_i \lambda^2} \left[\alpha' + \frac{1}{L} \ln \frac{1}{R}\right] + \frac{edg E_0}{\tau'} \left[\exp\left(\frac{kT}{E_0}\right) - 1\right] \qquad (2.157)$$

Using values of $\eta_i = 1$, $g = 5 \times 10^{18}$ cm^{-3} eV^{-1}, $d = 1\cdot5 \times 10^{-4}$ cm, $\tau' = 10^{-9}$ sec, $(\alpha' + (1/L) \ln (1/R)) = 50$ cm^{-1}, Dousmanis and coworkers were able to achieve remarkably good agreement with experimental results, but only at the expense of taking the value of E_0 as a temperature-dependent adjustable parameter. The value of d used here is, however, more realistic than that quoted on p. 54 from Lasher's work. This method is in contradiction with that of Unger (1967b) who gives the following approximate relation for E_0:

$$E_0 \simeq \frac{e^2}{\varepsilon} \frac{\sqrt{(4\pi N_I l)}}{2\cdot 457}$$

where N_I is the concentration of impurities and l is the screening length which is very nearly independent of temperature.

A less empirical method for calculating the temperature dependence of threshold currents, based on the emission rates dealt with in the previous section, was proposed by Lasher and Stern (1964). This technique assumes that the loss term α' in the threshold relation (2.144) is temperature independent; therefore the gain $-\alpha(E_m)$ and hence the peak value of the stimulated emission curve are both independent of temperature. Thus for a given value of the losses at a given level of excitation (specified by the positions of the quasi-Fermi levels) we may calculate the value of the threshold current from the total spontaneous emission using (2.149). The only remaining parameter is the variation of the excitation with temperature. Strictly, this should take into account the spatial distribution of carriers and the variation of the quasi-Fermi levels in the junction; however, the solution of such a problem for the degenerate junction at high injection levels is extremely complicated. Lasher and Stern's suggestion was to treat the recombination as occurring in homogeneously doped p-type material; it was therefore assumed that the hole population is large compared with the electron density and hence can be treated as broadly independent of temperature.

From page 38 the emission rates calculated by Lasher and Stern are given by (2.122) and (2.123) on the basis of a no $\mathbf{k}$-selection rule model. It is instructive to compare the deduced curves of threshold current density as a function of temperature (i.e. of $C^{-1} R_{\mathrm{sp}}$ as a function of T) with those resulting for the case of a rigid $\mathbf{k}$-selection rule, using Eqns. (2.120) and (2.121) and otherwise entirely similar assumptions as outlined above. Such a comparison is shown in Fig. 2.11. At high temperatures the Lasher–Stern curves tend to satisfy $j_t \propto T^\beta$, with β varying between 1·8 and 2·6. The corresponding curves with $\mathbf{k}$ selection tend to obey a 1·5-power law at high temperatures giving significantly lower values of threshold currents at room temperature. However, as mentioned on page 38, a rigid $\mathbf{k}$-selection rule is unlikely to be achieved for the junction laser

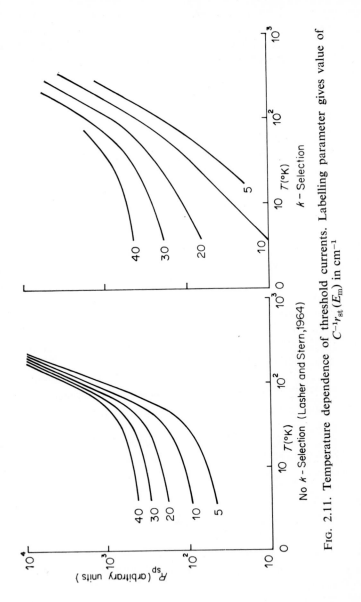

FIG. 2.11. Temperature dependence of threshold currents. Labelling parameter gives value of $C^{-1}r_{st}$ (E_m) in cm^{-1}

although the theory may apply to population inversions achieved by other means, e.g. electron bombardment or optical excitation.

It should be mentioned that an earlier calculation for the case of the k-selection rule (Hall, 1963) yielded the result of $j_t \propto T$ at high temperatures; however, the initial assumptions here were rather different. They included space-charge neutrality in the form $n = p$, i.e. it involved the placing of the active region at the point of the chemical junction where $N_D = N_A$. As we have seen above, this must now be regarded as somewhat unrealistic for the injection laser where the active region is usually in the p region. Another paper using the k-selection rule (Pikus, 1966) achieved the more usual T^3 variation for high temperatures by noting the increase of the penetration depth d with temperature.

An extension (Stern, 1966b) of the Lasher–Stern theory used the Gaussian band-tail density-of-states given in (2.125) and yielded better agreement with experiment in the form of: (a) a power-law variation of gain, $-\alpha(E_m)$, with current density, and (b) a T^3 variation of the threshold current at high temperatures. In addition it has been shown (Unger, 1967a, b) that the numerical results obtained by Lasher and Stern (1964) and Stern (1966b) may be approximated analytically.

2.5.6 A threshold calculation using exponential band tails

We conclude this section with a simple calculation (Adams, 1969) using the recombination model of Lasher and Stern together with the exponential densities of states given by (2.124):

$$g_c(E) = A_c \exp\left(\frac{E-E_c}{E_{0c}}\right), \quad g_v(E) = A_v \exp\left(\frac{E-E_v}{E_{0v}}\right) \quad (2.158)$$

Using these expressions in eqn. (2.117), one obtains

$$r_{st}(E) = B \int_{-\infty}^{\infty} g_c(E') g_v(E'-E) \left\{ \frac{1}{1+\exp\left[(E'-\mu_c)/kT\right]} \right.$$

$$\left. - \frac{1}{1+\exp\left[(E'-E-\mu_v)/kT\right]} \right\} dE'$$

$$= B A_c A_v kT\pi \, \mathrm{cosec}\left[kT\pi\left(\frac{1}{E_{0c}}-\frac{1}{E_{0v}}\right)\right]$$

$$\left\{ \exp\left[\left(\frac{E-E_G}{E_{0v}}\right)+(\mu_c-E_c)\left(\frac{1}{E_{0c}}-\frac{1}{E_{0v}}\right)\right] \right.$$

$$\left. - \exp\left[\left(\frac{E-E_G}{E_{0c}}\right)-(E_v-\mu_v)\left(\frac{1}{E_{0c}}-\frac{1}{E_{0v}}\right)\right] \right\} \quad (2.159)$$

where the factor B contains the matrix element for the transition and is slowly varying with photon energy, E. The convergence of the integrals appearing in this theory is subject to the conditions $E_{0c}, E_{0v} > kT$. Maximizing (2.159) and rewriting in terms of the gain, $-\alpha(E_m)$, one obtains

$$-\alpha(E_M) = \frac{\pi^2 c^2 \hbar^3}{\mu^2 E_M^2} BA_c A_v kT\pi \operatorname{cosec}\left[kT\pi\left(\frac{1}{E_{0c}} - \frac{1}{E_{0v}}\right)\right]$$

$$\times \left[\left(\frac{E_{0v}}{E_{0c}}\right)^{E_{0c}/(E_{0c}-E_{0v})} - \left(\frac{E_{0v}}{E_{0c}}\right)^{E_{0v}/(E_{0c}-E_{0v})}\right]$$

$$\times \exp\left[\frac{\mu_c - E_c}{E_{0c}} + \frac{E_v - \mu_v}{E_{0v}}\right] \qquad (2.160)$$

where

$$E_M = eV_a + \frac{E_{0c} E_{0v}}{E_{0c} - E_{0v}} \ln\left(\frac{E_{0v}}{E_{0c}}\right) \qquad (2.161)$$

Again, using the densities-of-states given in (2.158) one obtains for the total spontaneous emission rate

$$R_{sp} = Bpn$$

$$= BA_c A_v (kT\pi)^2 \operatorname{cosec}\left(\frac{kT\pi}{E_{0c}}\right) \operatorname{cosec}\left(\frac{kT\pi}{E_{0v}}\right)$$

$$\times \exp\left(\frac{\mu_c - E_c}{E_{0c}} + \frac{E_v - \mu_v}{E_{0v}}\right) \qquad (2.162)$$

Hence, using the threshold condition (2.144) and the expression (2.149) for the threshold current density and eliminating the exponential factors from (2.160) and (2.162), it follows that

$$j_t = \frac{-\alpha(E_M)\dfrac{ed\mu^2 E_M^2}{\eta_i \pi^2 c^2 \hbar^3} kT\pi \operatorname{cosec}\left(\dfrac{kT\pi}{E_{0c}}\right) \operatorname{cosec}\left(\dfrac{kT\pi}{E_{0v}}\right)}{\operatorname{cosec}\left[kT\pi\left(\dfrac{1}{E_{0c}} - \dfrac{1}{E_{0v}}\right)\right]\left[\left(\dfrac{E_{0v}}{E_{0c}}\right)^{E_{0c}/(E_{0c}-E_{0v})} - \left(\dfrac{E_{0v}}{E_{0c}}\right)^{E_{0v}/(E_{0c}-E_{0v})}\right]}$$

$$= \frac{-\alpha(E_M)\, ed\mu^2 E_M^2 \, kT\pi \left[\cot\left(\dfrac{kT\pi}{E_{0v}}\right) - \cot\left(\dfrac{kT\pi}{E_{0c}}\right)\right]}{\eta_i \pi^2 c^2 \hbar^3 \left[\left(\dfrac{E_{0v}}{E_{0c}}\right)^{E_{0c}/(E_{0c}-E_{0v})} - \left(\dfrac{E_{0v}}{E_{0c}}\right)^{E_{0v}/(E_{0c}-E_{0v})}\right]} \qquad (2.163)$$

This relation gives an approximately constant value of j_t for low temperatures. The curve then climbs steeply in the region where kT approaches E_{0c} or E_{0v}, whichever is the smaller. For temperatures above this value, i.e. about $120°K$, the theory breaks down. This is in accordance with the physical situation, since at these temperatures the quasi-Fermi levels will no longer be in the exponential tails and may have entered the parabolic parts of the bands. Hence above about $120°K$ the theory of Lasher and Stern (1964) or Stern (1966b) will be more applicable. These theories give the expected T^3 variation of the threshold current density at room temperature.

2.6 Electromagnetic theory of the injection laser

The use of the Fabry–Perot interferometer as a resonant cavity in the injection laser has already been noted above. In this case the term merely refers to a rectangular structure with two cleaved or polished ends a distance L apart. On pages 49–50 we have considered only the longitudinal propagation of a wave in the cavity. Such a wave is in fact confined to the vicinity of the junction by changes in the complex dielectric constant near the junction. These so-called guided modes are vital in ensuring that the wave propagates in the region where net gain occurs so that the threshold condition (2.144) may be satisfied at current values which are not unreasonably high.

In this section we will discuss further the longitudinal modes which determine the laser frequency, mode separation and their changes with temperature and other parameters. We also consider the field variation in a direction perpendicular to the junction with the object of estimating the diffraction losses, α_{diff}, and the penetration depth, d, of the wave away from the junction in terms of other parameters. In particular, some discussion is given of the nature of the dielectric discontinuities responsible for mode guiding.

2.6.1 Longitudinal (Fabry–Perot) modes

From Eqn. (2.142) we see that the Fabry–Perot cavity permits the formation of standing waves when its length contains an integral number, M, of half wavelengths:

$$L = \frac{M\lambda_M}{2\mu} \qquad (2.164)$$

where λ_M is the wavelength (in vacuum) and μ is some average value of the refractive index of the medium in the cavity. As explained in connection with (2.144), this result is applied to energies E_M near the maximum of the stimulated emission curve. For GaAs, with $\mu \simeq 3{\cdot}6$, $\lambda \simeq 8{,}500\,\text{Å}$ and $L \simeq 1$ mm one then finds $M \sim 10^4$.

The separation between wavelengths of adjacent modes (i.e. $\Delta M = -1$) is then (suppressing suffices M in the rest of this subsection):

$$\Delta\lambda = \frac{\lambda^2}{2L\left(\mu - \frac{\lambda}{\Delta\lambda}\frac{\Delta\mu}{}\right)} \qquad (2.165)$$

The factor $\frac{\Delta\mu}{\Delta\lambda}$ occurring in (2.165) may be approximated by $\left(\frac{\partial\mu}{\partial\lambda}\right)_T$, but cannot be neglected for semiconductor lasers although this is the case for most other solid-state lasers.

The wavelength of a mode may be altered by external influences such as pressure, uniaxial stress or a magnetic field. For example, the temperature dependence *of the wavelength* may be deduced as follows, from Eqn. (2.164):

$$\frac{d\lambda}{dT} = \frac{dL}{dT}\frac{2\mu}{M} + \frac{2L}{M}\left(\frac{\partial\mu}{\partial T}\right)_\lambda + \frac{2L}{M}\left(\frac{\partial\mu}{\partial\lambda}\right)_T\frac{d\lambda}{dT}$$

Since $\frac{1}{L}\frac{dL}{dT}$ is negligible for most semiconductors (including GaAs, where it is of order 10^{-6} per $^\circ$K), one obtains

$$\frac{1}{\lambda}\frac{d\lambda}{dT} = \frac{\frac{1}{\mu}\left(\frac{\partial\mu}{\partial T}\right)_\lambda}{\left[1 - \frac{\lambda}{\mu}\left(\frac{\partial\mu}{\partial\lambda}\right)_T\right]} \qquad (2.166)$$

From experiment (Burns and Nathan, 1964), in GaAs, the variation of λ with T is 0.046 Å/$^\circ$K; theoretical values predicted by (2.166) are 50% higher than this. Combining Eqns. (2.165) and (2.166) one obtains an expression for the rate of change of refractive index with temperature in terms of the mode separation $\Delta\lambda$:

$$\left(\frac{\partial\mu}{\partial T}\right)_\lambda = \frac{\lambda}{2L\Delta\lambda}\left(\frac{d\lambda}{dT}\right) \qquad (2.167)$$

Alternatively, one may write Eqns. (2.165) and (2.166) *in terms of the frequency v*, rather than the wavelength

$$\Delta v = \frac{c}{2L\left[\mu + v\frac{\Delta\mu}{\Delta v}\right]} \qquad (2.168)$$

$$-\frac{dv}{dT} = \frac{v\left(\frac{\partial\mu}{\partial T}\right)_v}{\left[\mu + v\left(\frac{\partial\mu}{\partial v}\right)_T\right]} \qquad (2.169)$$

Equation (2.169) corresponds to (2.166) and again implies a condition which experimental values should satisfy.

In order to recast these results, note that

$$\left(\frac{\partial \mu}{\partial T}\right)_\nu = -\left(\frac{\partial \mu}{\partial \nu}\right)_T \left(\frac{\partial \nu}{\partial T}\right)_\mu$$

If the change in μ is due merely to a shift in the absorption edge, one obtains:

$$\left(\frac{\partial \mu}{\partial T}\right)_\nu = -\left(\frac{\partial \mu}{\partial \nu}\right)_T \frac{1}{h} \frac{dE_G}{dT} \tag{2.170}$$

Hence, using (2.168), (2.169) and (2.170), one finds a simple relation between energy gap, laser frequency and separation between adjacent modes:

$$\frac{d(h\nu)}{dT} = \left[1 - \frac{2L\mu\Delta\nu}{c}\right] \frac{dE_G}{dT} \tag{2.171}$$

Equation (2.171) may be used to predict the variation of laser frequency with temperature, using experimental values of $\Delta\nu$ and $\frac{dE_G}{dT}$; it gives this variation to within 5% of the observed values (Engeler and Garfinkel, 1963).

2.6.2 Transverse variation of the field

For the purposes of an electromagnetic description, the idealized model of a p–n junction laser consists of a thin active region where an inverted population exists, bounded by two thick passive regions. As mentioned above, the modes in such a structure are confined in the vicinity of the junction by discontinuities in the complex dielectric constant. The waveguide effect which results is responsible for the near- and far-field patterns of the laser radiation. It is appropriate at this point to discuss the nature of the changes in the dielectric constant as one passes from one region to the neighbouring region. Early calculations (Lasher, 1963; Hall and Olechna, 1963; McWhorter and coworkers, 1963) on this effect assumed a change in only the imaginary part of the dielectric constant. This resulted in values of the diffraction loss, α_{diff} (mentioned in Section 2.5), which were much greater than those found from experiment. Some authors have considered waveguiding due to a change only in the real part of the dielectric constant (Yariv and Leite, 1963, 1964) whilst others have used a more general three-layer slab model (McWhorter, 1963; Stern, 1965; Anderson, 1965).

2.6.3   Mechanisms for the dielectric discontinuity

In general, the dielectric constants in the three regions may be written as

$$\varepsilon_i = \varepsilon_i' + i\varepsilon_i'' \qquad (i = 1, 2, 3) \tag{2.172}$$

We are here anticipating the notation of Fig. 2.12 in which the indices 1, 2, 3 refer to the p-type region, the active region and the n-type region, respectively. Various possible reasons for the discontinuities in the dielectric constant have been proposed. One of these assumes the real part discontinuity to be the result of free-carrier absorption in the n and p regions, characterized by plasma frequencies ω_{pn} and ω_{pp}, respectively. Here the first subscript stands for 'plasma', and

$$\omega_{pn}^2 = \frac{4\pi n e^2}{m_c}; \quad \omega_{pp}^2 = \frac{4\pi p e^2}{m_v}$$

The value of the dielectric constant in the n region is then lower than that in the active region by ω_{pn}^2/ω^2, where ω is the frequency of emitted radiation; similarly the dielectric constant in the p region is lower by a factor ω_{pp}^2/ω^2. For typical GaAs diodes, this yields discontinuities of about 6×10^{-4} (Yariv and Leite, 1963, 1964).

A more general method of estimating these discontinuities is furnished by the dispersion relations which connect the real and imaginary parts of the dielectric constant. These relations, often termed the Kramers–Kronig relations,[7] are a special case of the general interrelationships of the real and imaginary parts of an analytic function (the Hilbert transform relations), and may be written, in this case, as:

$$\left. \begin{aligned} \varepsilon'(\omega) - 1 &= \frac{2}{\pi} \mathscr{P} \int_0^\infty \frac{\omega' \varepsilon''(\omega')}{\omega'^2 - \omega^2} \, d\omega' \\ \varepsilon''(\omega) &= \frac{2\omega}{\pi} \mathscr{P} \int_0^\infty \frac{\varepsilon'(\omega')}{\omega^2 - \omega'^2} \, d\omega' \end{aligned} \right\} \tag{2.173}$$

where $\mathscr{P}$ denotes the Cauchy principal value of the integrals.

More significantly, the first of these relations may be applied to the function:

$$\sqrt{\mu_0 \varepsilon} = \mu + i\kappa, \tag{2.174}$$

yielding:

$$\mu(\omega) - 1 = \frac{2}{\pi} \mathscr{P} \int_0^\infty \frac{\omega' \kappa(\omega') \, d\omega'}{\omega'^2 - \omega^2},$$

---

[7] See, for instance, Stern (1963).

or in terms of the absorption coefficient, α, and photon energy, $E = \hbar\omega$,

$$\mu(E) - 1 = \frac{\hbar}{\pi} c\mathscr{P} \int_0^\infty \frac{\alpha(E')\, dE'}{E'^2 - E^2} \tag{2.175}$$

Thus one may estimate the magnitude of the change, $\Delta\mu_n$, in refractive index between the depletion region and the n region of a laser diode:

$$\Delta\mu_n = \frac{\hbar c}{\pi} \mathscr{P} \int_0^\infty \frac{\alpha(E') - \alpha_n(E')\, dE'}{E'^2 - E^2} \tag{2.176}$$

Similarly, for the change between the depletion region and the p region:

$$\Delta\mu_p = \frac{\hbar c}{\pi} \mathscr{P} \int_0^\infty \frac{\alpha_p(E') - \alpha(E')}{E'^2 - E^2}\, dE' \tag{2.177}$$

Here $\alpha(E')$ is the absorption coefficient for the active region which may be estimated from the stimulated emission calculations of Section 2.4. Similarly, α_p and α_n may be written as analytic functions of photon energy and other junction parameters. In particular, the following expression (Moss, 1959) for a simple absorption curve with an exponential edge has been used (Anderson, 1965):

$$\alpha_{n,p}(E') = \left\{ \frac{\alpha_\infty}{1 + \exp[\beta(E_1 - E')]} \right\}_{n,p} \tag{2.178}$$

where α_∞ is of order $10^4\,\mathrm{cm}^{-1}$ for GaAs; values of E_1 and β are available from experimental data on absorption. Thus values of the quantities $\Delta\mu_n, \Delta\mu_p$ may be deduced from equations (2.176) and (2.177) as a function of photon energy E.

To use the calculated values of the $\Delta\mu$'s and $\Delta\alpha$'s to estimate the corresponding changes $\Delta\varepsilon', \Delta\varepsilon''$ in the real and imaginary parts of the dielectric constant, note that from eqn. (2.174),

$$\sqrt{\mu_0[\varepsilon' + \Delta\varepsilon' + i(\varepsilon'' + \Delta\varepsilon'')]} = \mu + \Delta\mu + i(\kappa + \Delta\kappa)$$

Hence to first order

$$\Delta\varepsilon' \simeq \frac{2\Delta\mu\sqrt{\varepsilon'}}{\sqrt{\mu_0}} \tag{2.179}$$

and

$$\Delta\varepsilon'' \simeq \frac{c\Delta\alpha\sqrt{\varepsilon'}}{\omega\sqrt{\mu_0}} \tag{2.180}$$

where the assumptions $\varepsilon' \gg \Delta\varepsilon'$, $\Delta\varepsilon''$, ε'' have been used. In all cases it is found (Anderson, 1965) that the discontinuity in the imaginary part of ε is a second-order effect and that reasonably good results may be obtained by considering only a real part discontinuity and using Eqns. (2.176) to (2.179) to estimate its magnitude.

2.6.4 Solution of Maxwell's equations

Using our previous notation, Maxwell's equations may be written:[8]

$$\text{div}\,\mathscr{E}_i = 0, \qquad\qquad \text{div}\,\mathbf{H}_i = 0$$

$$\text{curl}\,\mathscr{E}_i = -\frac{\mu_0}{c}\frac{\partial \mathbf{H}_i}{\partial t'}, \quad \text{curl}\,\mathbf{H}_i = \frac{\varepsilon_i}{c}\frac{\partial \mathscr{E}_i}{\partial t'}. \tag{2.181}$$

Eliminating $\mathbf{H}_i$ in the usual way, we obtain the wave equation

$$\nabla^2 \mathscr{E}_i - \frac{\varepsilon_i \mu_0}{c^2}\frac{\partial^2 \mathscr{E}_i}{\partial t'^2} = 0 \tag{2.182}$$

A similar equation for $\mathbf{H}_i$ is obtained by eliminating $\mathscr{E}_i$ from (2.181). The wave equation (2.182) will now be applied to each of the three regions in the idealized electromagnetic model of a laser diode shown in Fig. 2.12. Also shown in the figure is the assumed variation of the electric and magnetic fields with the x coordinate. As in Section 2.5, the z axis is taken along the direction perpendicular to the mirrors and the waves are assumed to propagate in this direction as $\exp(iqz + i\omega t')$; i.e. the assumed solutions of the wave equation in the three regions are:

$$\left.\begin{aligned}
\mathscr{E}_1 &= \mathscr{E}_0\, A e^{i(p_1 x + qz + \omega t')} \\
\mathscr{E}_2 &= \mathscr{E}_0 \cos(p_2 x + \phi)\, e^{i(qz + \omega t')} \\
\mathscr{E}_3 &= \mathscr{E}_0\, B e^{i(-p_3 x + qz + \omega t')}
\end{aligned}\right\} \tag{2.183}$$

In this simple model, the y-variation of the field is neglected since the width of the device in this direction greatly exceeds the thickness, t, of the active region. However, in a recent paper Unger (1967c) considers a waveguiding effect occurring also in the y-direction as a result, e.g. of inhomogeneities in the junction doping. This may be responsible for the phenomenon of 'filaments' seen in the operation of most lasers (Jonscher and Boyle, 1967). Unger's paper gives reasonably simple expressions for the loss on the basis of a model whose absorption coefficient and refractive index vary smoothly in space.

Substituting the assumed solutions (2.183) into equation (2.182), one obtains:

$$p_i^2 + q^2 - \frac{\mu_0 \varepsilon_i \omega^2}{c^2} = 0 \qquad (i = 1, 2, 3) \tag{2.184}$$

---

[8] Here t' denotes time to distinguish it from the thickness t, introduced later.

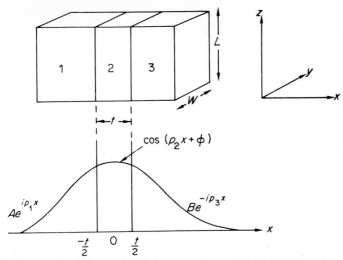

FIG. 2.12. Schematic representation of the x-dependent part of the electric field in the electromagnetic wave travelling along the z axis

At this point we must impose the boundary conditions at the interfaces of the three regions of the model, i.e. at $x = \pm t/2$ in Fig. 2.12. There are two classes of waves to be considered; those polarized with the electric field vector lying in the plane of the slab (the y–z plane) and those with the magnetic field vector lying in this plane. These are termed transverse electric (TE) and transverse magnetic (TM) waves, respectively. In both cases the boundary conditions imply continuity of the transverse electric and magnetic fields at $x = \pm t/2$. Thus, for the case of TE waves, one obtains:

$$Ae^{-ip_1t/2} = \cos\left(\phi - \frac{p_2 t}{2}\right)$$

$$ip_1 Ae^{-ip_1t/2} = -p_2 \sin\left(\phi - \frac{p_2 t}{2}\right)$$

$$Be^{-ip_3t/2} = \cos\left(\phi + \frac{p_2 t}{2}\right)$$

$$-ip_3 Be^{-ip_3t/2} = -p_2 \sin\left(\phi + \frac{p_2 t}{2}\right)$$

Hence

$$\tan\left(\phi - \frac{p_2 t}{2}\right) = -\frac{ip_1}{p_2}$$

and

$$\tan\left(\phi + \frac{p_2 t}{2}\right) = \frac{ip_3}{p_2}$$

It follows that

$$\tan(p_2 t) = \frac{ip_2(p_1 + p_3)}{p_2^2 + p_1 p_3} \tag{2.185}$$

Similarly, for the TM waves, the boundary conditions take the form

$$\tan(p_2 t) = i \frac{\left(\dfrac{p_2}{\varepsilon_2}\right)\left(\dfrac{p_1}{\varepsilon_1} + \dfrac{p_3}{\varepsilon_3}\right)}{\left[\left(\dfrac{p_2}{\varepsilon_2}\right)^2 + \left(\dfrac{p_1}{\varepsilon_1}\right)\left(\dfrac{p_3}{\varepsilon_3}\right)\right]} \tag{2.186}$$

Thus for each case (TE and TM) there are four equations, (2.184) and (2.185) or (2.186), to be solved to give p_i ($i = 1, 2, 3$) and q in terms of the other parameters. Several authors have solved these equations approximately by using the small argument expansion of $\tan(p_2 t)$ in (2.185) and (2.186). However, Anderson (1965) has shown by detailed numerical solutions that the small argument expansion is frequently not valid for GaAs lasers, especially in cases where the change in the dielectric constant is not symmetrical about the active region. The remaining parameter whose value is to a large extent uncertain is the active region width, t. This has been variously estimated as 0.055μ (Yariv and Leite, 1963, 1964), 3.3μ (Anderson, 1965) and 2μ (Burns and Nathan, 1964); it seems likely that these last two values are of the right order. It is then possible to completely determine theoretically the fields in the three regions, by evaluating the constants A, B, ϕ appearing in the assumed solutions (2.183) of Maxwell's equations.

2.6.5 Attenuation (gain) of a guided wave

The time-averaged power dissipation, P_i, per unit length and width in the three regions of Fig. 2.12 may be written[9] as follows:

$$P_1 = \frac{\omega\varepsilon''}{8\pi}\int_{-\infty}^{-t/2} |\mathscr{E}_1|^2 dx = -\frac{\omega\varepsilon''}{8\pi}\frac{|\mathscr{E}_0|^2 A^2 e^{p_1 t}}{2p_1''} \quad (p_1'' < 0) \tag{2.187a}$$

$$P_2 = \frac{\omega\varepsilon''}{8\pi}\int_{-t/2}^{t/2} |\mathscr{E}_2|^2 dx = \frac{\omega\varepsilon''}{8\pi}\left|\frac{\mathscr{E}_0}{2}\right|^2 \left[\frac{\cos 2\phi \sin p_2' t}{p_2'} + \frac{\sinh p_2'' t}{p_2''}\right] \tag{2.187b}$$

$$P_3 = \frac{\omega\varepsilon''}{8\pi}\int_{t/2}^{\infty} |\mathscr{E}_3|^2 dx = -\frac{\omega\varepsilon''}{8\pi}\frac{|\mathscr{E}_0|^2 B^2 e^{p_3 t}}{2p_3''} \quad (p_3'' < 0) \tag{2.187c}$$

---

[9] See, for example, Landau and Lifshitz (1960).

where we have used the notation: $p_i = p_i' + ip_i''$. Similarly, the time-averaged energy stored in the three regions is:

$$U_1 = \frac{\varepsilon'}{8\pi} \int_{-\infty}^{-t/2} |\mathscr{E}_1|^2 \, dx = -\frac{\varepsilon'}{8\pi} \frac{|\mathscr{E}_0|^2 A^2 e^{p_1't}}{2p_1''} \qquad (p_1'' < 0) \qquad (2.188a)$$

$$U_2 = \frac{\varepsilon'}{8\pi} \int_{-t/2}^{t/2} |\mathscr{E}_2|^2 \, dx = \frac{\varepsilon'}{8\pi} \left|\frac{\mathscr{E}_0}{2}\right|^2 \left[\frac{\cos 2\phi \sin p_2' t}{p_2'} + \frac{\sinh p_2'' t}{p_2''}\right] \qquad (2.188b)$$

$$U_3 = \frac{\varepsilon'}{8\pi} \int_{t/2}^{\infty} |\mathscr{E}_3|^2 \, dx = -\frac{\varepsilon'}{8\pi} \frac{|\mathscr{E}_0|^2 B^2 e^{p_3''t}}{2p_3''} \qquad (p_3'' < 0) \qquad (2.188c)$$

Now, using Eqn. (2.174) and expanding the square root to first order in $(\varepsilon''/\varepsilon')^2$, one may obtain the following relation:

$$\kappa \simeq \frac{\sqrt{\mu_0}\, \varepsilon''}{2\sqrt{\varepsilon'}}$$

or, in terms of the absorption coefficient, α,

$$\alpha \simeq \frac{\omega \sqrt{\mu_0}\, \varepsilon''}{c\sqrt{\varepsilon'}} \qquad (2.189)$$

It follows, using the approximation $\varepsilon' \simeq \varepsilon_1', \varepsilon_2', \varepsilon_3'$, that the following simple expression (Anderson, 1965) describes the attenuation (or gain) of a guided wave:

$$\alpha = \left(\frac{P_1 + P_2 + P_3}{U_1 + U_2 + U_3}\right) \frac{\sqrt{\mu_0}\, \varepsilon'}{c} \qquad (2.190)$$

This quantity is what was termed $[\alpha(E_M) + \alpha_0 + \alpha_{\text{diff}}]$ in Section 2.5 and we may therefore use it to estimate the diffraction losses. Alternatively, it has been used (Anderson, 1965) to rederive the threshold condition (2.144) in a slightly different form.

Also, from the assumed solutions of Maxwell's equations, we may estimate values of the penetration depth, d, of the electromagnetic wave away from the active region, using the approximate relation:

$$d \simeq \left[\frac{1}{|p_1''|} + \frac{1}{|p_3''|}\right] \qquad (2.191)$$

Using a similar relation to (2.191) together with the assumption that the change in dielectric constant is due to free carrier effects, Yariv and Leite (1963, 1964) obtain the extremely high value of $d \simeq 45\mu$. In view of the lack of exact knowledge of many parameters in the electromagnetic theory of the injection laser, it is frequently the case that experimental values of d are used and the theory inverted to yield estimates of the active region thickness, t.

2.6.6 Angular divergence of the injection laser beam

The light emitted from an injection laser is never found to have the highly directional properties of other laser beams. Basically, this is due to the very small dimensions of the active region of the device. The angle of divergence of the beam in a given direction varies inversely as the dimension of the active region in that direction, e.g. perpendicular to the plane of the junction, $\theta \simeq \lambda/t$. Thus, using the coordinate system of Fig. 2.12, it is clear that this angle will be much greater in the x–z plane than that in the y–z plane.

An accurate description of the angular distribution in the x–z plane requires a complete solution of the electromagnetic mode problem outlined above and characterized by eqns. (2.184), (2.185) and (2.186). However, in the y–z plane the values of the angular divergence at which maxima in the light output occur may be estimated (Burns and coworkers, 1965) as a simple consequence of the laws of reflection and refraction. This will now be described, even though the method has not been entirely successful. These maxima are assumed to be the result of modes in the form of standing waves characterized by a number N' (as in Fig. 2.13) which gives the ratio of the number of reflections at the ends of the cavity to the number at the sides. Thus one obtains (with the notation of the figure):

$$\tan \theta_i = \frac{W}{N'L},$$

$$\sin \theta_0 = \mu \sin \theta_i$$

i.e.

$$\sin \theta_0 = \mu \sin \left[\tan^{-1} \left(\frac{W}{N'L} \right) \right] = \frac{\mu\, W/N'L}{\sqrt{[1+(W/N'L)^2]}} \qquad (2.192)$$

This shows that θ_0 decreases with the quantity $x \equiv W/N'L$. From Fig. 2.13 one would expect in any case a decrease in the angle of divergence of the beam if N' or L are increased at fixed W. Equation (2.192) has been used to infer the mode type (N') from experiment.

2.6.7 External quantum efficiency and laser output

Although only indirectly associated with the electromagnetic theory of the injection laser, we conclude this section with a derivation of simple expressions for the external quantum efficiency, output power and power conversion efficiency. We distinguish first between the internal and external quantum efficiencies of stimulated radiation, η_i and η_{ext}, respectively, defined as follows:

η_i = number of photons produced inside the cavity per injected carrier,

η_{ext} = number of photons emitted from the cavity per injected carrier.

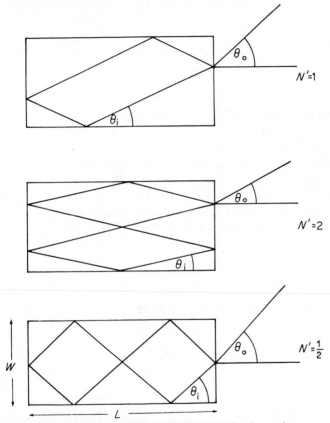

FIG. 2.13. Some standing wave modes in the cavity

It follows that:

$$\eta_{\text{ext}} = \eta_i \times \frac{\text{number of photons emitted from cavity per injected carrier}}{\text{number of photons produced inside cavity per injected carrier}}$$

This equation applies above threshold to photons produced both by spontaneous and stimulated emission. However, when applied to stimulated emission only, the external quantum efficiency will be denoted by $\Delta\eta_{\text{ext}}$.

Reference to Eqn. (2.144) shows that at or above the threshold this latter ratio for stimulated emission only may be simply expressed as (Biard and coworkers, 1964)

$$\frac{\frac{1}{L}\ln\frac{1}{R}}{\alpha' + \frac{1}{L}\ln\frac{1}{R}}$$

i.e.

$$\Delta\eta_{\text{ext}} = \frac{\eta_i}{\frac{L\alpha'}{\ln\frac{1}{R}} + 1} \tag{2.193}$$

This has been termed an incremental efficiency. More complicated expressions for $\Delta\eta_{\text{ext}}$ have been derived but are basically of the same form as (2.193) (Cheroff and coworkers, 1963).

To determine the output power of an injection laser in a steady-state (C.W.) situation, we assume that the threshold current density, j_t, is 'wasted' in producing spontaneous emission, heating effects, etc., and that all the remaining current density $(j - j_t)$ is used in producing coherent output. Thus with an applied voltage V_a, for a laser of length L and width W, one obtains the following result for the output power P:

$$P = \Delta\eta_{\text{ext}}(j - j_t)LWV_a \tag{2.194}$$

From (2.194) it follows that the total power conversion efficiency, η, of an injection laser may be expressed as:

$$\eta = \frac{\Delta\eta_{\text{ext}}(j - j_t)LWV_a}{jLWV_a}$$

$$= \Delta\eta_{\text{ext}}\left(1 - \frac{j_t}{j}\right) \tag{2.195}$$

Using the derived results for $\Delta\eta_{\text{ext}}$ and j_t, (2.195) and (2.151), respectively, it is then possible to maximize η with respect to the parameter L. This yields typical values of $L \simeq 10^{-2}$ cm (Akselrad, 1966) in qualitative agreement with experiment; however, more detailed results are unlikely in view of the uncertainty of the other parameters in the theory.

2.7 General discussion

In conclusion, it may be convenient to survey the broad requirements for a good semiconductor laser.

2.7.1 Direct material

On page 27 and subsequently, attention was confined to direct materials; GaAs falls into this category and no reliable report of lasing from indirect materials is available. One may write on a rough kinetic picture for the radiative transition rate

$$R_{\mathrm{d}} = B_{\mathrm{d}} n_{\mathrm{d}} p \quad \text{and} \quad R_{\mathrm{ind}} = B_{\mathrm{ind}} n_{\mathrm{ind}} p \qquad (2.196\mathrm{a, b})$$

These refer to transitions from a direct minimum (electron concentration n_{d}) and from an indirect minimum, respectively. One knowns that B_{ind} is in general much smaller that B_{d}; the numerical values are of order 10^{-15} cm^3/sec and 10^{-8} cm^3/sec, respectively, for germanium (Haynes and Nilsson, 1965). Thus any given electron concentration in the lowest minimum is in a sense wasted if the minimum is indirect. However, (2.196a, b) presuppose that there is no **k**-selection rule, and indeed the integrated form of Eqn. (2.123) assumes the form (2.196a) or (2.196b) in a sufficiently simple case of this kind. If a **k**-selection rule holds, Eqn. (2.121) applies, and the number of states which can contribute to photon emission of a given energy is strictly limited. This is a more complicated situation and Eqns. (2.196a, b) no longer hold. In this case the indirect transition can be expected to contribute more effectively to the emitted radiation, because the partial lifting of the **k**-selection rule due to phonon participation can bring more electron states into action. In spite of this, however, the direct materials are clearly the best to use, since for them $n_{\mathrm{d}} \gg n_{\mathrm{ind}}$ supports the inequality $B_{\mathrm{d}} \gg B_{\mathrm{ind}}$.

The above discussion applies to spontaneous emission only (the photon occupation number would be needed in (2.196a, b) for stimulated emission; see Eqns. (2.81)). However, the quantum-mechanical factors for stimulated emission are the same as the corresponding ones for spontaneous emission (see Eqn. (2.86)). In this sense what has been said about the difference between direct and indirect transitions applies also to stimulated emission. This difference was first noted by Dumke (1962).

2.7.2 Degenerate conditions

The condition for negative absorption has been stated in a number of equivalent ways at the end of Section 2.3. For band–band transitions an equation like (2.95) implies that the quasi-Fermi levels have to be separated by at least the energy gap. This means that at least one band must be degenerate. Usually, of course, both bands are degenerate. Thinking in terms of a more general spectrum of electron states, and possibly impurity states near the band edges, the condition requires that at least one quasi-Fermi level must lie in a spectrum of active states.

2.7.3   Heavy doping

The need for heavy doping is a consequence of the degeneracy requirement. The doping produces band tails and the quasi-Fermi levels for the active region lie within these tails. In this way a good density-of-states is produced for the radiative transitions and a partial or complete suspension of the **k**-selection rule is produced which further increases the density of electron states which participate in the transition. These points are additional to the advantage which arises from producing band tails, which reduce the separation which is needed between quasi-Fermi levels to bring about population inversion (Winogradoff and Kessler, 1964, 1966, 1967).

Quantitative models for the description of the requirements 2.7.1 to 2.7.3 are contained in Sections 2.3 and 2.4, the basic quantum mechanics being introduced in Section 2.2. We now turn to more specific requirements.

2.7.4   Uniformity of junction

Light emission from filaments in a laser structure is usual and may be due to inhomogeneities. The existence of localized regions of high current densities is detrimental in leading to a loss of efficiency due to more pronounced heating effects than would be found in a uniform material. Recently dislocation-free GaAs has been prepared in which fuller use is made of the junction (Hatz, 1967).

2.7.5   Laser quality factor Q_M

To ask for a large laser quality factor is fully equivalent to asking for a small value of the quantity (2.144):

$$-\alpha(E_m) = \alpha'(E_m) + (1/L)\ln(1/R) = (\mu/c\tau_M) = \mu E_M/c\hbar Q_M,$$

or for a long photon lifetime τ_M in the dominant laser modes. These objects can be achieved in the obvious way by keeping down the losses α' or maximizing the reflectivity R (with one of the two reflectivities equal to unity). An increase in length L achieves the same effect, but for given threshold current density j_t, it increases the current input $j_t LW$ required, and this is often undesirable. In fact, as is clear from the discussion on pages 70–72, there are optimum values for L and R to give the best power conversion efficiency which also determine the threshold current.

2.7.6   Laser geometry and threshold current density j_t

Cylindrical, triangular and other geometries have been used in injection lasers, but have so far proved to be less important than the type with two parallel reflecting surfaces discussed here. For this type $j_t \propto -\alpha(E_m)$

(Eqn. (2.151)) so that conditions favouring large Q_M, as discussed under 2.7.5, also favour a low threshold current density. In fact, there is good experimental confirmation of the semiempirical law $j_t = a + b/L$, where a and b are constants (Pilkuhn and coworkers, 1965).

Reduction of j_t for given current density j also increases the power output and the power conversion efficiency (Eqns. (2.188) and (2.189)). The latter can be maximized by an optimal choice of L and R.

2.7.7 Narrow spontaneous linewidth

To see the need of a narrow spontaneous linewidth, one should imagine R_{sp} in the 'demerit' factor γ to be expressed as an integral over photon energy which is then approximated by the rectangle made up from the maximum of the integrand and the linewidth $\Delta E = h\Delta v$. It is then clear that ΔE cancels out of γ, and from (2.151) that j_t can be kept down by keeping Δv small. Physically this means that one minimizes power wastage in frequencies away from the spontaneous emission maximum in the neighbourhood of which lasing modes develop, as explained in our discussion of Eqn. (2.147).

The achievement of a narrow spontaneous linewidth is, however, more difficult in an injection laser than it is for a laser based on sharp atomic levels. The low energy rise of the line is governed by the rise in the density of states and the high energy decline is determined by the spread of the occupied states about the Fermi level. Thus Δv decreases with temperature, but it is unlikely that really small values of Δv can be obtained if a *band* of allowed states is involved.

2.7.8 Temperature dependence of threshold current

The demerit factor is near unity at $T = 0$ and, like Δv, it increases with temperature. Equation (2.151) thus shows that the threshold current rises with temperature. One can look at this effect in a different way, which should, however, be equivalent. In passing from the threshold at $T = 0$, $\gamma = 1$, to the threshold at elevated temperatures one must fill additional states to energy kT over the Fermi level due to the thermal broadening of the electron occupation probability. An additional current density is therefore required. This leads to an experimental T^3-dependence of j_t at elevated temperatures, and the whole curve can be understood theoretically best by dropping the parabolic density-of-states in favour of an exponential or Gaussian distribution of states.

The quantity $-\alpha(E_m)$ is called the gain and is usually proportional to threshold current density, $-\alpha(E_m) = \beta j_t$, where β is the gain factor. The temperature dependence of j_t and of $1/\beta$ are similar in shape.

2.7.9   Other factors

There is an obvious need to keep down the penetration depth d of the electromagnetic wave since this represents a loss of energy from the cavity. This can be achieved by adequate changes in dielectric constant to produce waveguide effects. It is found that the diffraction losses are lowest for the case of dielectric symmetry about the active region. Anderson (1965) has calculated that no mode-guiding takes place below the critical active region width of about $3\cdot3\mu$ for a GaAs laser at room temperature. Also of interest in connection with the electromagnetic problem is the result of Hatz and Mohn (1967) that TM modes have lower thresholds than TE modes when the real and imaginary parts of the dielectric constant are assumed to vary linearly with distance in the active region; this polarization is indeed found experimentally in diodes made from dislocation-free GaAs.

The internal quantum efficiency should be as high as possible with a minimum of non-radiative transitions such as recombination by phonon cascades or Auger effects. Although a theoretical treatment is still outstanding, it appears that doping requirements to obtain maximum efficiency for a given material at a given temperature can be determined experimentally (Dobson, 1966).

The influence of laser width on laser performance has so far received little theoretical attention but has been found experimentally (Pilkuhn, 1968) to have some effect on thresholds and efficiencies. This may be due to the formation of transverse electromagnetic modes or possibly modes of total internal reflection. Another factor which may perhaps affect performance as a function of width is the heating effect at high current densities. The consideration of such factors as heat flow and series resistance yields optimization of parameters such as laser area and mirror reflectivities for the C.W. case (Vilms and coworkers, 1966) and current values and pulse lengths for pulsed operation (Keyes, 1965).

References

Adams, M. J. (1969). *Solid State Electron.* **12**, 661
Akselrad, A. (1966). *Appl. Phys. Letters*, **8**, 250.
Anderson, W. W. (1965). *J. Quantum Electronics*, **QE-1**, 228.
Bethe, H. A. (1964). *Intermediate Quantum Mechanics*, Benjamin, New York.
Bernard, M. G. A. and B. Duraffourg (1961). *Phys. Stat. Sol.*, **1**, 699.
Biard, J. R., W. N. Carr and B. S. Reed (1964). *Trans. AIME*, **230**, 286.
Birnbaum, G. (1964). *Optical Masers*, Academic Press, New York.
Bohm, D. (1951). *Quantum Theory*, Prentice-Hall, New York.
Boyd, G. D. and J. P. Gordon (1961). *Bell System Tech. J.*, **40**, 489.

Burns, G., R. A. Laff, S. E. Blum, F. H. Dill, Jnr. and M. I. Nathan (1963). *IBM J. Res. Develop.*, **7**, 62.
Burns, G., and M. I. Nathan (1964). *Proc. I.E.E.E.*, **52**, 770.
Calloway, J. (1963). *J. Phys. Chem. Solids*, **24**, 1063.
Cheroff, G., F. Stern and S. Triebwasser (1963). *Appl. Phys. Letters*, **2**, 173.
Cohen, M. E. and P. T. Landsberg (1967). *Phys. Rev.*, **154**, 683.
Dirac, P. A. M. (1927). *Proc. Roy. Soc.*, **A114**, 243.
Dobson, C. D. (1966). *Brit. J. Appl. Phys.*, **17**, 187.
Dousmanis, G. C., H. Nelson and D. L. Staebler (1964). *Appl. Phys. Letters*, **5**, 174.
Dumke, W. P. (1962). *Phys. Rev.*, **127**, 1559.
Dumke, W. P. (1963). *Phys. Rev.*, **132**, 1998.
Eagles, D. M. (1960). *J. Phys. Chem. Solids*, **16**, 76.
Edwards, S. F. (1961). *Phil. Mag.*, **6**, 617.
Einstein, A. (1917). *Phys. Zeit.*, **18**, 127.
Engeler, W. E. and M. Garfinkel (1963). *J. Appl. Phys.*, **34**, 2746.
Evans, D. A. and P. T. Landsberg (1963). *Solid State Electron.*, **6**, 169.
Fox, A. G. and T. Li (1961). *Bell Syst. Tech. J.*, **40**, 453.
Frisch, H. L. and S. P. Lloyd (1960). *Phys. Rev.*, **120**, 1175.
Haken, H. (1965). In F. Sauter (Ed.), *Festkörperprobleme*, Vol. IV. Vieweg, Braunschweig. p. 1.
Hall, R. N., G. E. Fenner, J. D. Kingsley, T. J. Soltys and R. O. Carlson (1962). *Phys. Rev. Letters*, **9**, 366.
Hall, R. N. (1963). *Solid State Electron.*, **6**, 405.
Hall, R. N., and D. J. Olechna (1963). *J. Appl. Phys.*, **34**, 2565.
Halperin, B. I., and M. Lax (1966a). *Phys. Rev.*, **148**, 722.
Halperin, B. I., and M. Lax (1966b). *Phys. Rev.*, **153**, 802.
Hatz, J. (1967). *J. Quantum Electronics*, **QE-3**, 643.
Hatz, J., and E. Mohn (1967). *J. Quantum Electronics*, **QE-3**, 656.
Haynes, J. R., and N. G. Nilsson (1965). In *Radiative Recombination in Semiconductors*, Dunod, Paris. p. 21.
Heitler, W. (1954). *Quantum Theory of Radiation*, 3rd ed. Clarendon Press, Oxford.
Ivey, H. F. (1966). *J. Quantum Electronics*, **QE-2**, 713.
Jones, R., and T. Lukes (1969). *Proc. Roy. Soc. A*, **309**, 457.
Jonscher, A. K., and M. H. Boyle (1967). In *Proceedings of the International Symposium on Gallium Arsenide, Reading*, 1966, I.P.P.S., London. p. 78.
Kane, E. O. (1957). *J. Phys. Chem. Solids*, **1**, 249.
Kane, E. O. (1963a). *Phys. Rev.*, **131**, 79.
Kane, E. O. (1963b). *Phys. Rev.*, **131**, 1532.
Kane, E. O. (1966). In R. K. Willardson and A. C. Beer (Eds.), *Physics of III–V Compounds*, Vol. 1. Academic Press, New York. p. 75.
Keyes, R. W. (1965). *IBM J. Res. Develop.*, **9**, 303.
Klauder, J. R. (1961). *Ann. Phys. (N.Y.)*, **14**, 43.
Knox, R. S. (1963). *Theory of Excitons*, Academic Press, New York.
Kohn, W. (1957). In F. Seitz and D. Turnbull (Eds.), *Solid State Physics*, Vol. 5. Academic Press, New York, p. 257.
Kotik, J., and M. C. Newstein (1961). *J. Appl. Phys.*, **32**, 178.

Lamorte, M. F., S. Caplan and T. Gonda (1967). *In Proceedings of the International Symposium on Gallium Arsenide, Reading*, 1966. I.P.P.S., London, p. 88.

Landau, L. D., and E. M. Lifshitz (1960). *Electrodynamics of Continuous Media*, Pergamon Press, London.

Landsberg, P. T. (1966). *Phys. Stat. Sol.*, **15**, 623.

Landsberg, P. T. (1967a). In O. Madelung (Ed.), *Festkörperprobleme*, Vol. VI. Vieweg, Braunschweig. p. 174.

Landsberg, P. T. (1967b). *Phys. Stat. Sol.*, **19**, 777.

Landsberg, P. T. (1967c). *Solid State Electron.*, **10**, 513.

Landsberg, P. T. (1969). In *Solid State Theory: Methods and Applications*, John Wiley and Sons, New York.

Lasher, G. J. (1963). *IBM J. Res. Develop.*, **7**, 58.

Lasher, G. J., and F. Stern (1964). *Phys. Rev.*, **133**, A553.

Lax, B. (1964). In C. H. Townes and P. A. Miles (Eds.), *Quantum Electronics and Coherent Light*, Academic Press, New York. p. 17.

Lax, M. (1952). *Phys. Rev.*, **85**, 621.

Lax, M., and J. C. Phillips (1958). *Phys. Rev.*, **110**, 41.

Leite R. C. C., J. C. Sarace, D. H. Olson, B. G. Cohen, J. M. Whelan and A. Yariv (1965). *Phys. Rev.*, **137**, A1583.

Lengyel, B. A. (1966). *Introduction to Laser Physics*, John Wiley and Sons, New York.

Lucovsky, G. (1965). *Solid State Commun.*, **3**, 105.

Lucovsky, G. (1966). In P. L. Kelley, B. Lax and P. E. Tannenwald (Eds.), *Physics of Quantum Electronics*, McGraw-Hill, New York. p. 467.

Marinelli, F. (1965). *Solid State Electron.* **8**, 939.

Mashkevich, V. S. (1967). *Laser Kinetics*, Elsevier, New York.

Mayburg, S. (1963). *J. Appl. Phys.*, **34**, 1791.

McWhorter, A. L. (1963). *Solid State Electron.*, **6**, 417.

McWhorter, A. L., H. J. Zeiger and B. Lax (1963). *J. Appl. Phys.*, **34**, 235.

Moll, J. L., and J. F. Gibbons (1963). *IBM J. Res. Develop.*, **7**, 157.

Morgan, D. J., and P. T. Landsberg (1965). *Proc. Phys. Soc.*, **86**, 261.

Morgan, D. J., and J. A. Galloway (1967). *Phys. Stat. Sol.*, **23**, 97.

Morgan, T. N. (1965). *Phys. Rev.*, **139**, A294.

Moss, T. S. (1959). *Optical Properties of Semiconductors*, Butterworths, London.

Nathan, M. I., W. P. Dumke, G. Burns, F. H. Dill Jnr. and G. J. Lasher (1962). *Appl. Phys. Letters*, **1**, 62.

Nathan, M. I. (1966). *Appl. Opt.*, **5**, 1514, and *Proc. I.E.E.E.*, **54**, 1276.

Nelson, D. F., M. Gershenzon, A. Ashkin, L. A. d'Asaro and J. C. Sarace (1963). *Appl. Phys. Letters*, **2**, 182.

Pankove, J. I. (1965). *Phys. Rev.*, **140**, A2059.

Parmenter, R. H. (1955). *Phys. Rev.*, **97**, 587.

Parmenter, R. H. (1956). *Phys. Rev.*, **104**, 22.

Pátek, K. (1967). *Lasers*, Iliffe, London.

Pikus, G. E. (1965). *Fiz. Tverd. Tela.*, **7**, 3536 (*Soviet Phys. Solid State*, **7**, 2854, 1966).

Pilkuhn, M. H., H. Rupprecht and J. Woodall (1965). *J. Quantum Electronics*, **QE-I**, 184.

Pilkuhn, M. H. (1968). *Phys. Stat. Sol.*, **25**, 9.
Popov, Yu. M. (1967). *Appl. Opt.*, **6**, 1818.
Quist, T. M., R. H. Rediker, R. J. Keyes, W. E. Krag, B. Lax, A. L. McWhorter and H. J. Zeiger (1962). *Appl. Phys. Letters*, **1**, 91.
Sah, C. T., R. N. Noyce and W. Shockley (1957). *Proc. I.R.E.*, **45**, 1228.
Schawlow, A. L., and C. H. Townes (1958). *Phys. Rev.*, **112**, 1940.
Stern, F. (1963). In F. Seitz and D. Turnbull (Eds.), *Solid State Physics*, Vol. 15, Academic Press, New York. p. 299.
Stern, F. (1965). In *Radiative Recombination in Semiconductors*, Dunod, Paris. p. 165.
Stern, F. (1966a). In R. K. Willardson and A. C. Beer (Eds.), *Physics of III–V Compounds*, Vol. 2. Academic Press, New York. p. 371.
Stern, F. (1966b). *Phys. Rev.*, **148**, 186.
Unger, K. (1965). *Fortschr. Phys.*, **13**, 701.
Unger, K. (1967a). *Z. Physik.*, **207**, 322.
Unger, K. (1967b). *Z. Physik.*, **207**, 332.
Unger, K. (1967c). *Ann. Phys. (Lpz.)*, **19**, 64.
Vilms, J., L. Wandinger and K. L. Klohn (1966). *J. Quantum Electronics*, **QE-2**, 80.
Vinetskii, V. L., V. S. Mashkevich and P. M. Tomchuk (1964). *Fiz. Tverd. Tela.*, **6**, 2037. (*Soviet Phys. Solid State*, **6**, 1607, 1965.)
Wilson, A. H. (1953). *Theory of Metals*, 2nd. ed. Cambridge University Press, Cambridge.
Wilson, D. K. (1963). *Appl. Phys. Letters*, **3**, 127.
Winogradoff, N. N., and H. K. Kessler (1964). *Solid State Commun.*, **2**, 119.
Winogradoff, N. N., and H. K. Kessler (1966). *Int. J. Electronics*, **21**, 329.
Winogradoff, N. N., and H. K. Kessler (1967). *Solid State Commun.*, **5**, 155.
Wolff, P. A. (1962). *Phys. Rev.*, **126**, 405.
Yariv, A., and J. P. Gordon (1963). *Proc. I.E.E.E.*, **51**, 4.
Yariv, A., and R. C. C. Leite (1963). *Appl. Phys. Letters*, **2**, 55.
Yariv, A., and R. C. C. Leite (1964). In P. Grivet and N. Bloembergen (Eds.), *Proceedings of the Third International Congress on Quantum Electronics*, Columbia University Press, New York. p. 1879.
Zeiger, H. J. (1964). *J. Appl. Phys.*, **35**, 1657.

3

The Properties of Gallium Arsenide
p-n Junction Lasers

C. H. GOOCH

Services Electronics Research Laboratory, Baldock, Hertfordshire

Contents

3.1   Introduction

The gallium arsenide lasers with which this book is concerned are *p–n* junction devices. They are operated at forward bias and under suitable conditions emit an intense beam of radiation at a wavelength of approximately 0·9 μ.

A schematic diagram of a typical device is shown in Fig. 3.1 and its electrical characteristics in Fig. 3.2. At low current densities spontaneous radiation is produced isotropically in the junction region. Some of this radiation will be reflected back into the junction region by the polished faces of the device and at sufficiently high current densities this radiation will be amplified by stimulated transitions. When the amplification is sufficient to overcome the losses in the system the device will oscillate and an intense beam of stimulated radiation will emerge from the device at the intersection of the junction plane with the cavity faces.

In general, devices must be operated at high current densities and this frequently means that they must be operated under pulsed conditions in order to keep the power dissipation in the device down to an acceptable level. Pulsed operation produces a variety of thermal effects in a laser so that the observed properties of a laser are to some extent modified by the manner in which they are used.

In this chapter a description will be given of the somewhat idealized properties which can be observed under carefully controlled experimental conditions. It will also be shown how the observed properties depart from these when lasers are operated under conditions such that thermal effects in the device become important.

In describing device properties an attempt will also be made to distinguish between laboratory experiments using carefully selected devices in order to obtain the best possible results, and the results which are more likely to be obtained using commercially available devices. This distinction arises because semiconductor lasers are still at a relatively early stage of development with a wide gulf between the 'best' and 'typical' devices. This of course arises with any new device and one hopes that, as development continues, results which can only be obtained in the laboratory at present will be obtainable on typical and commercially available devices in the future.

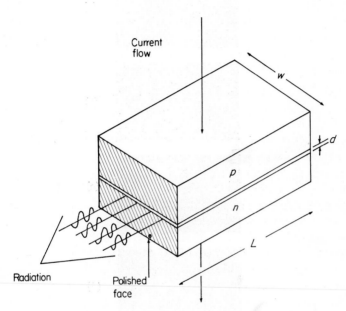

FIG. 3.1. Schematic diagram of a semiconductor laser

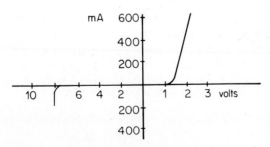

FIG. 3.2. Electrical characteristics of a semiconductor laser

3.2 Threshold properties

3.2.1 *Determination of threshold current*

The radiation from a gallium arsenide laser occurs in the near infrared region of the spectrum. Visual observations of this radiation can be made

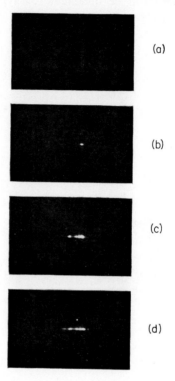

0·5mm

FIG. 3.3. The appearance of a laser junction at
successively higher current densities.
(a) 1000 A/cm²   (c) 1500 A/cm²
(b) 1250 A/cm²   (d) 1750 A/cm².
The onset of laser action in (b) can be seen

using a suitable image converter and the radiation can be detected using a suitable photomultiplier or silicon junction detector.

The appearance of a laser junction as the current density through it is increased is shown in Fig. 3.3. At low current densities the junction can be seen as a uniform line source of spontaneous radiation. As the current is increased a clearly defined point is reached at which the brightness of small regions of the junction increases markedly (Broom and coworkers, 1963). This critical current can be defined as the threshold current of the laser although this definition is somewhat subjective and there are more precise ways of specifying this threshold point. As the current density is further increased more of these bright regions occur until it is difficult to resolve any individual bright regions and the junction appears to be emitting uniformly. However, even at the highest current densities, observation at high magnification reveals considerable structure in the appearance of the junction.

If both ends of the laser are examined it is seen that the two emission patterns (frequently referred to as 'near field patterns') are mirror images of each other. It thus appears that laser action is occurring in filaments along the junction plane and perpendicular to the polished faces of the cavity.

It is believed that this non-uniform emission is due to defects or inhomogeneity in the device. If the junction is not flat, or variations in doping level occur in the junction region, a lasing filament will pass through regions of low gain and this will increase the threshold of that filament. In an ideal device laser action would occur simultaneously across the whole width of the junction and in fact the most efficient devices appear to approach this ideal condition.

Measurements of the power emitted from a laser as the current through it is increased are shown in Fig. 3.4. Below a critical current the output is small. Above this current the output increases rapidly and linearly with current and the threshold current of the laser can be defined as the current at which this break occurs. It is found that the threshold defined in this way agrees very closely with that obtained from visual observations.

3.2.2 Threshold current and cavity length

It is expected that the current density required to achieve laser action in a device would depend on the length of the laser cavity and the reflectivity of the cavity ends. It was shown in Chapter 2 that, at a given temperature, the threshold condition for a laser is given by

$$R \exp(gL) = \exp(\alpha L) \tag{3.1}$$

or

$$g = \alpha + \frac{1}{L} \ln \frac{1}{R} \tag{3.2}$$

4

where L is the cavity length, R is the reflectivity of the cavity ends, α is the loss per unit length in the cavity and g is the gain per unit length in the cavity and will be a function of the current density through the device.

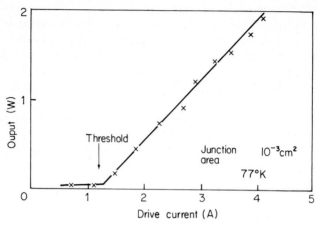

Fig. 3.4. The power output from a laser as a function of current

Pilkuhn and Rupprecht (1963) have studied this relationship by measuring the threshold current of a series of lasers which differed only in their cavity lengths. They found that Eqn. (3.2) was valid and that the gain was a linear function of the current density of the device, thus

$$g(J_0) = \beta J_0 = \alpha + \frac{1}{L}\ln\frac{1}{R} \qquad (3.3)$$

where J_0 is the threshold current density and β is a constant at a given temperature.

Some typical measurements of J_0 versus $1/L$ from diffused junction lasers are shown in Fig. 3.5. For these particular results, obtained at $77°K$, $\alpha = 15$ cm^{-1} and $\beta = 2 \times 10^{-2}$ cm/A. These values correspond to current densities $\sim 10^3$ A/cm^2 so that to make devices which will operate at reasonable currents the junction areas must be small. In practice devices can be made with junction areas of 10^{-4} to 10^{-2} cm^2 and threshold currents from 0·1 to 10 A at $77°K$.

Techniques for determining the gain and loss coefficients, β and α, have been of considerable value in studying the properties of different laser structures at different temperatures and further use of the results of such experiments will be made in subsequent sections of this chapter.

3.2.3   The effects of temperature and structure on threshold current

The details of the ways in which one can make *p–n* junctions suitable for lasers are considered in detail in another chapter. However, to understand some of the threshold current properties of devices it is necessary to bear in mind the different ways in which the junctions are made and the basic distinction between diffused and epitaxial junctions are outlined here.

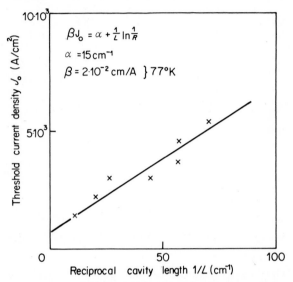

$$\beta J_0 = \alpha + \frac{1}{L} \ln \frac{1}{R}$$

$$\alpha = 15 \, cm^{-1}$$

$$\beta = 2 \cdot 10^{-2} \, cm/A \;\} \, 77°K$$

FIG. 3.5. The variation of laser threshold current density with cavity length

[i] *Diffused junctions.*   Diffused junctions can be made by diffusing an acceptor such as zinc into an *n*-type material; a schematic diagram of the resulting impurity profile is shown in Fig. 3.6a. This is probably the most controllable method of making laser junctions and is entirely suitable for devices intended to operate at 77°K.

[ii] *Epitaxial junctions.*   An epitaxial junction is made by growing a layer of *p*- (or *n*-) type material on a substrate of *n*- (or *p*-) type material. A schematic diagram of the resulting impurity profile is shown in Fig. 3.6b; this profile is somewhat idealized as some diffusion inevitably occurs during the epitaxial process and will distort the profiles shown. As a result of this diffusion the *p–n* junction will be formed within the original substrate material and the *p*-type material adjacent to the junction will be compensated to some extent. There is some evidence that this compensation is a necessary requirement for efficient laser action.

Winogradoff and Kessler (1964) and Dousmanis and coworkers (1964) have shown that diodes made using epitaxial techniques can have lower threshold currents at room temperature than diffused junction diodes.

The threshold current of a typical laser is approximately 10^3 A/cm² at 77°K although values as low as 500 A/cm² have been observed in well-constructed devices. This threshold current increases approximately as the

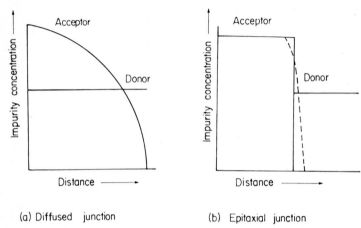

(a) Diffused   junction   (b) Epitaxial junction

Fig. 3.6. Impurity profiles in diffused and epitaxial junctions

cube of the absolute temperature so that, at room temperature, values $\sim 5 \times 10^4$ A/cm² are obtained (Fig. 3.7). Below 77°K the variation is less rapid. In calculating the performance of devices operating at, or above, 77°K sufficient accuracy is obtained by assuming a cubic variation so that the threshold current I' at a temperature ΔT above the ambient T_0 is

$$I' = I_0(1 + \Delta T/T)^3 \tag{3.4}$$

Dobson (1966) has measured the threshold current at 77°K of diffused lasers made from material with various doping levels and found an optimum at 2×10^{18} cm⁻³. However, it is not clear whether this represented a true optimum or whether material of higher doping level had a poorer crystalline perfection and thus resulted in poorer lasers.

Considerable attempts have been made to obtain devices with low threshold currents at room temperature. By careful control of the diffusion process, threshold current densities as low as 3.6×10^4 A/cm² can be obtained at 300°K (Carlson, 1966) but lower values than this appear to be dependent on the use of epitaxial processes. By using such techniques it is possible to obtain threshold currents $\sim 3 \times 10^4$ A/cm² (Nelson and coworkers, 1964; Susaki, 1967). However, to obtain a useful power output from a laser it must be operated at a current considerably in excess of the threshold current so that a small reduction in threshold current is of less importance than a high efficiency above threshold. A comparison of the threshold current versus temperature characteristics of typical diffused and epitaxial lasers is shown in Fig. 3.8.

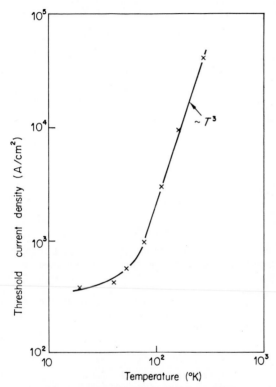

FIG. 3.7. The threshold current of a diffused laser as a function of temperature showing a T^3 dependence above 80° K

Pilkuhn, Rupprecht and Woodall (1965) have determined the gain and loss coefficients of a series of diffused and epitaxial lasers at liquid nitrogen and room temperatures and obtained the values of α and β given in Table 3.1.

It can be seen that, at 77°K, the devices are similar but at room temperature the gain coefficient of the epitaxial device is several times greater than that of the diffused devices. These results also show that it is the decrease in β with temperature which is the major cause of the increase in threshold current. The increase in α with temperature is insignificant in

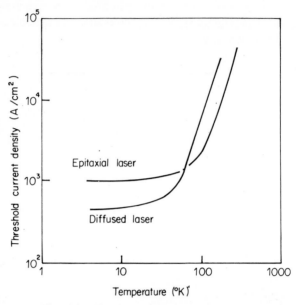

Fig. 3.8. The threshold current of diffused and epitaxial lasers as a function of temperature

Table 3.1. Typical gain and loss coefficients for diffused and epitaxial lasers.

	Temperature (°K)	α (cm^{-1})	β (cm/A)
Diffused lasers	77	15	$2 \cdot 5 \times 10^{-2}$
	296	20	$5 \cdot 7 \times 10^{-4}$
Epitaxial lasers	77	14	$3 \cdot 9 \times 10^{-2}$
	296	92	$3 \cdot 8 \times 10^{-3}$

the case of the diffused devices and is the smaller of two factors in epitaxial devices.

In general the threshold properties of diffused and epitaxial devices are in good agreement with calculations of the gain in a laser made by Stern (1966) (Chapter 2). Stern showed that lower threshold current densities at room temperature would be obtained using junctions in very heavily doped material and epitaxial techniques provide the means of obtaining the required high doping levels. However, a detailed comparison of experimental results with theoretical predictions is difficult since the exact impurity levels and profiles in epitaxial junctions are hard to determine.

3.3 Power output and efficiency

3.3.1 Definitions

The term 'efficiency' has been applied to semiconductor lasers in several ambiguous ways so that to avoid confusion it is necessary to define and explain the terms which are in common use.

In general, a laser will emit radiation from both the polished ends of the cavity. It is usually possible to make use of the radiation emitted from only one end of the device so that in this case half the device output is wasted. This can be overcome if one end of the cavity is coated to increase its reflectivity and if this end of the cavity has a reflectivity of 100% all the emitted radiation will emerge from one end of the device. In quoting the power emitted from a device care must be taken to specify whether it is the useful power emitted from one end of the device or the total power from both ends which is given. Since in the general case the cavity ends can have arbitrary reflectivities with a different, non-zero, output from either end it is convenient to refer always to the total output from a device. In a laser in which one end has a reflectivity of 100% this is also useful output from a device.

[i] *Quantum efficiency* η. The quantum efficiency of an electro-luminescent device can be defined as the ratio of the number of photons emitted by the device to the number of electrons which flow through the device. In general the energy of the emitted photon need not be the same as that of the electron so that the quantum efficiency of a device need not be the same as the power efficiency of the junction. However, in a semi-conductor laser the forward voltage, V, across the junction when forward biased is approximately equal to the band gap E_G and the energy of the emitted photon is approximately E_G.

Hence the device quantum efficiency is given by

$$\eta = \frac{P/h\nu}{I/e} = \frac{P}{IV} \tag{3.5}$$

where P is the power emitted, $h\nu$ is the photon energy, V is the junction voltage, I is the current through the device and e is the electronic charge.

It has been shown (Fig. 3.4) that below threshold the output from a laser is small but rises linearly with current above threshold. The quantum efficiency of the laser is thus effectively zero below threshold but increases as the current is increased above threshold, approaching a limiting value at high currents. The quantum efficiency of the device is thus a function of its operating current and this is clearly inconvenient if one wishes to specify the properties of a device.

[ii] *Incremental quantum efficiency $\Delta\eta$*. Since the output of a laser increases linearly with current above threshold we can define an incremental quantum efficiency which is independent of the current at which the device is operating.

Restricting this definition to operating points above threshold

$$\Delta\eta = \frac{P}{(I-I_0)\,V} \tag{3.6}$$

and $\Delta\eta$ is now a constant for a given device at a particular temperature.

The incremental quantum efficiency of gallium arsenide lasers is high, values of 50% at 77°K and 30% at room temperature being obtainable although of course the efficiency of the average device will be somewhat lower than these figures.

[iii] *Power efficiency η_P*. In the discussion of efficiency we have so far ignored the series electrical resistance, R, of a device. In most cases this resistance is appreciable and to calculate the power efficiency of a device one must take into account power dissipated in this resistance.

The total power dissipated in a device is

$$W = I_0\,V + (I-I_0)(1-\Delta\eta)\,V + I^2\,R \tag{3.7}$$

In most cases of practical interest the device will be operated at a current considerably above the threshold current and if $\Delta\eta < 1$ we have, with sufficient accuracy,

$$W = IV + I^2\,R \tag{3.8}$$

and

$$P = (I-I_0)\,V\,\Delta\eta \tag{3.9}$$

The power efficiency of the device is

$$\eta_P = \frac{(I-I_0)\,V\,\Delta\eta}{IV + I^2\,R} \tag{3.10}$$

$$\approx \frac{\Delta\eta}{1 + IR/V} \quad \text{if } I \gg I_0$$

The resistive dissipation becomes more important at high currents and is most significant in devices operating at room temperature.

To illustrate the various points which have been made in this discussion of device efficiency we have the example of data for a typical device operating at 77 or 300°K given in Table 3.2.

This example illustrates the general point that although comparable peak powers can be obtained from devices operating at liquid nitrogen or room temperatures the power efficiency of the device differs by an order of magnitude. We will also see in a later section that the mean power obtainable from a device is severely limited at room temperature.

TABLE 3.2. Basic properties of a typical laser operating at 77 and 300°K

	77°K	300°K	
Operating temperature			
Threshold current	1	50	A
Series resistance	0·1	0·1	Ω
Incremental quantum efficiency	30	15	%
Operating current	20	100	A
Output power (peak)	9	11	W
Input power (peak)	70	1150	W
Power efficiency	13	1	%

3.3.2 Efficiency–temperature effects

In the numerical example given in the previous section it was indicated that the incremental quantum efficiency of a laser does not decrease drastically as the operating temperature is increased. Thus is further illustrated in Fig. 3.9 showing the output of a laser as a function of current at a series of temperatures between 77 and 300°K.

Although $\Delta\eta$ can be relatively insensitive to temperature the power efficiency of a device decreases rapidly with temperature since at high temperatures devices must be operated at high current densities in order to exceed the threshold current. In such cases the dissipation in a device is largely the Joule heating in the series resistance and in the development of devices for operation at room temperature this resistance must be kept to a minimum.

3.3.3 Radiation patterns

One of the outstanding characteristics of a gas laser is the small divergence of the beam of radiation emitted. This characteristic is not shared by semiconductor lasers. If the devices were perfect then the radiation pattern would be that of a coherent source of dimensions $d \times w$ where d is the thickness of the emitting region and is approximately $2\,\mu$ and w is the width of the cavity. The diffraction limited beam from a laser 0·1 cm wide would thus have a divergence of $25 \times 0·05°$ approximately.

In practice this condition is not observed since the effective aperture of a typical device is by no means a source of coherent radiation. A typical radiation pattern of a laser is shown in Fig. 3.10 and is seen to consist of a complex family of lobes. This is frequently referred to as the 'far field pattern' of the laser. The details of these patterns vary considerably from

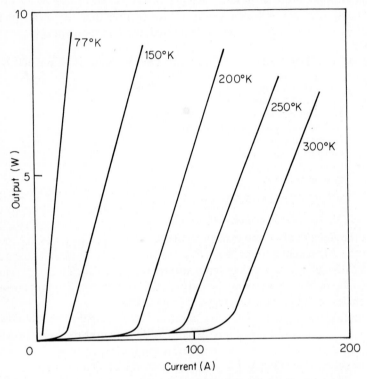

Fig. 3.9. The output from a laser as a function of current, at different temperatures

laser to laser and attempts to interpret the details on the basis of a model of the laser structure have met with limited success.

Antonoff (1964) and Ludman and Hergenrother (1966) have shown that the structure perpendicular to the junction plane can be interpreted as being due to different modes propagating in the laser cavity. Radiation from different modes emerges from the laser in different directions giving a family of beams whose angular spacing depends on the cavity length and changes in the dielectric constant within the structure. Dyment (1967) has shown that specially constructed lasers with a junction width of 50 μ can

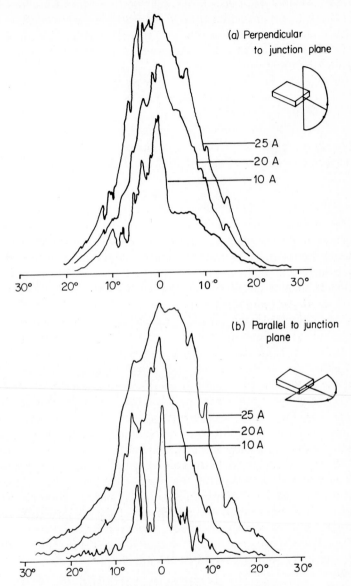

FIG. 3.10. Radiation patterns from a laser junction at various currents (threshold current 5 A, junction width 0·5 mm, 77° K) (after Seany, 1967)

exhibit radiation patterns which can be described by a Hermite–Gaussian function. Extensions of this work by Dyment and Zachos (1968) on devices with an even narrower junction ($w = 13\ \mu$, $L = 400\ \mu$) have shown that it is possible to obtain simple Gaussian intensity distributions in the junction plane for current values up to 60% above threshold.

In general, however, it has not been possible to interpret in detail the radiation patterns which are obtained when lasers are operated well above threshold. In practice it can be seen from Fig. 3.10 that to make effective use of the radiation emitted by a junction laser it must be collimated with an optical system with an aperture of approximately $f/2$. The optical system will then collect 90% of the radiation emitted by the laser.

3.3.4   The theory of laser efficiency

In a forward biased semiconductor p–n junction minority carriers are injected across the junction and recombine with the majority carriers. If the junction is made in a direct band gap semiconductor the probability that this recombination will be a radiative process is high. In a gallium arsenide junction this probability approaches 100% at 77°K and may be as high as 50% at room temperature.

Below threshold the radiative recombination in a laser diode is a spontaneous process and the photons which are produced are emitted isotropically. These photons can be absorbed in the heavily doped material on either side of the junction and, of the photons which do reach the surface of the material, only a small fraction will emerge due to the high refractive index of gallium arsenide and the correspondingly small critical angle. In gallium arsenide, with a refractive index of 3·5, the critical angle is approximately 17° so that only radiation incident on a surface at angles less than this can emerge from the material.

Above threshold the radiative recombination is a stimulated process and the resulting photons travel in the plane of the junction perpendicular to the polished faces of the laser. In this case the amount of radiation which is emitted is determined by the reflectivity of the gallium arsenide–air or gallium arsenide–vacuum interface. With a refractive index of 3·5 this reflectivity is approximately 30% so that 70% of the radiation incident on the face of a laser cavity will be transmitted.

Within the cavity some photons will be lost from the system, either by diffraction out of the narrow active region of the device or by free carrier absorption within the active region and it has been shown that these losses can be represented by a loss coefficient, α. This is the loss coefficient which occurred in the laser threshold relationship in Section 3.2.

Biard and coworkers (1964) have given the following formula for the incremental quantum efficiency of a laser

$$\Delta\eta = \frac{\ln 1/R}{\alpha L + \ln 1/R} \qquad (3.11)$$

This formula applies to a device in which the reflectivity of either end is the same, but it can be extended to the case in which the reflectivity of one end of the device is increased to 100%.

Gooch (1966) has proposed a model of a laser by means of which one can calculate the power emitted by either end of a cavity of arbitrary reflectivities. Unfortunately this model does not give rise to a simple formula for the efficiency of a device and results must be obtained from numerical solutions to a differential equation. Both models show that the efficiency of a device will decrease as its length is increased and give identical results under conditions where both models are applicable.

The predicted efficiency of devices as a function of cavity length is shown in Fig. 3.11. These curves are plotted for values of α which are typical of lasers operating at 77 and 300°K and apply to a laser with a reflectivity of 30% at either end. If one end of the device has a reflectivity of 100%, its efficiency will be the same as that of a symmetrical laser of double the cavity length and its efficiency can be obtained from these same curves.

From these curves we can see that a 'silvered' laser will have a lower efficiency than an unsilvered laser of the same cavity length. However, since all the power from a silvered laser emerges from one end of the device the useful power from a laser is increased considerably by 'silvering'.

When the reflectivity of one end of a laser is increased the threshold current of the device will be decreased. 'Silvering' is thus a useful technique in two respects and is commonly applied to all practical devices.

The efficiencies which are predicted by (3.11) and shown in Fig. 3.11 are likely to be optimistic since two assumptions have been made: (*a*) All the transitions which occur in a laser are radiative transitions. As has been indicated, this is probably a valid assumption at low temperatures but may be in error by a considerable factor at room temperature. (*b*) Laser action occurs uniformly over the whole area of the junction. In general this is true only in the best devices so that any calculation of efficiency is the maximum value which could be expected for a cavity of given length.

3.3.5 *Device design*

The brightness of a source of radiation can be defined as the power emitted per unit source area into a unit solid angle and this has the units

98   *C. H. Gooch*

of W/cm² steradian or the equivalent. In many applications it is this
brightness which is of fundamental importance. In general any optical
system used to collimate the radiation from a laser will not change the
apparent brightness of the source, since if a smaller image of the source is
produced the radiation will be emitted into a correspondingly larger solid
angle thus maintaining a constant brightness.

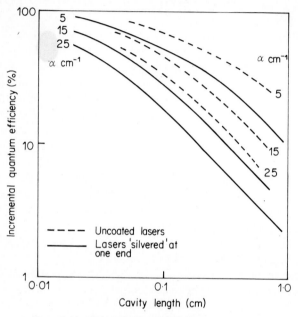

FIG. 3.11. The calculated efficiency of a laser as a
function of cavity length

In any application one will in general be reliant on devices which are
commercially available. However, in an ideal situation one would optimize
the device design with the system requirements bearing in mind such
factors as peak and mean power required, the size of the optical system,
amplitude of the current pulse which can be produced with adequate
rise-time, etc.

Considering a device of unit width, the brightness of the source is
given by

$$B = \frac{P}{d\omega} = \frac{(J-J_0)LV\Delta\eta}{d\omega}$$ (3.12)

where d is the width of the emitting region, ω is the solid angle into which
radiation is emitted and the other symbols have been previously defined.

Little control over d or ω is possible and it is shown elsewhere that the emitting region has a width $\sim 2\,\mu$ and that radiation is emitted into a cone of semi-angle approximately 15° thus requiring an $f/2$ optical system.

Various conditions might be imposed upon J, the current density at which the device is operated and L the cavity length. For example, to avoid overheating it may be necessary to stipulate a maximum value of J and in

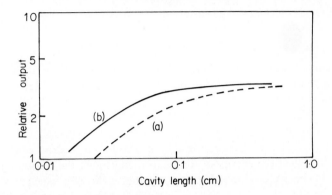

FIG. 3.12. The relative power output available from devices operated at constant current density, as a function of cavity length. (a) unsilvered devices, (b) one cavity end with 100% reflectivity

this case the total current through the device will be JL per unit width. Although the power emitted by the device is given by $(J-J_0)LV\Delta\eta$ there is no point in increasing L indefinitely since, as the length increases, $L.\Delta\eta$ approaches a constant value given by $\ln(1/R)/\alpha$ and the brightness, B, saturates (Fig. 3.12). In practice it appears that the optimum cavity length in this case is about 0·2 cm, or 0·1 cm if one end of the cavity is silvered.

In the previous paragraph it was assumed that the current required to drive the laser would be available. However, in many applications of lasers this will not be true. This is particularly so if the laser is to be operated by current pulses since the production of current pulses with a fast rise-time presents considerable problems. In this case it is obviously necessary to restrict the total area of the laser and, if the width of the laser is as small as is practicable, the cavity length must be restricted. Restrictions such as these are commonly encountered when devices are operated at room temperature where the current densities required to reach the threshold condition are high. In this case small devices are common and cavities with dimensions as small as $50\,\mu$ wide and $100\,\mu$

long have been made. A laser of this area would have a threshold current of less than 5 A at room temperature.

The points which have been made in this section emphasize the general principle that in order to obtain the optimum performance from any system involving the application of gallium arsenide lasers, the laser should be designed with the requirements of the particular system in mind. This, of course, would be the ideal situation and in practice one must usually design a system to make the best use of the devices available.

3.4   Spectral properties

In a semiconductor laser radiation is produced by the recombination of an excess minority carrier with a majority carrier. In a gallium arsenide junction laser electrons are injected into the conduction band of the *p*-type material and recombine with the majority carriers, i.e. holes, in the valence band. We will consider in more detail the nature of the energy levels involved in this transition in a later section, but it can be seen that one would expect the energy of the photon emitted in the recombination process to be approximately equal to the band gap energy of the semiconductor.

In gallium arsenide the band gap energy is approximately 1·4 eV and the emitted radiation has a wavelength of approximately 8500 Å. However, by using other semiconductors with a suitable direct band structure a wide range of wavelengths can be obtained. A list of the materials which have been used to make junction lasers is given in Table 3.3. The range of wavelengths which has been obtained in this way extends from 6400 Å, using an alloy of gallium arsenide with gallium phosphide, to 14 μ using a lead–tin telluride alloy.

3.4.1   Spontaneous emission

When a diode is operated at a current below the threshold value, radiation is produced by spontaneous recombination in the vicinity of the junction and is emitted isotropically. Typical spectra of a diode operating in this way are shown in Fig. 3.13. As is to be expected, the total light output from the diode increases as the current through it increases. It can also be seen that, at the lower temperatures, the peak of the emission shows a shift to shorter wavelengths, or higher energy, as the current increases.

The shift of spectrum has been interpreted by Nelson and coworkers (1963) on a model which is known as the 'band filling model'. According to this, the increasing density of minority carriers injected into the *p*-type region of the diode must fill the conduction band to increasingly higher

TABLE 3.3. Semiconductors which have been used as p–n junction laser materials.[a]

Material		Wavelength (μ)	Material		Wavelength (μ)
Gallium arsenide	GaAs	0.85	Gallium arsenide–phosphide	$GaAs_xP_{1-x}$	0.64–0.85
Indium phosphide	InP	0.91	Indium–gallium arsenide	$In_xGa_{1-x}As$	0.83
Gallium antimonide	GaSb	1.5	Indium arsenide–phosphide	$InAs_xP_{1-x}$	0.89
Indium arsenide	InAs	3.1	Gallium–aluminium arsenide	$Ga_xAl_{1-x}As$	0.7
Indium antimonide	InSb	5.4			
Lead sulphide	PbS	4.3	Lead–tin telluride	$Pb_xSn_{1-x}Te$	9.4–13.7
Lead telluride	PbTe	6.5	Lead–tin selenide	$Pb_xSn_{1-x}Se$	10.2–12.4
Lead selenide	PbSe	8.5			

[a] The emission wavelength is approximate, and the values quoted are those observed at low temperatures.

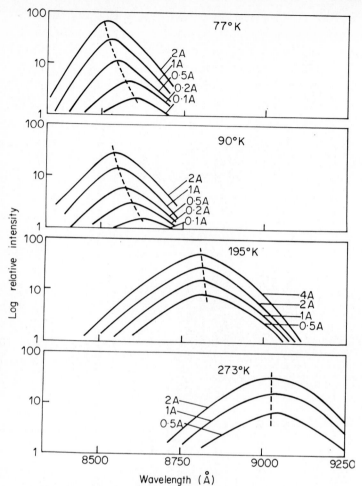

FIG. 3.13. The spontaneous emission spectrum of a diode at various temperatures and current densities (diode area 1.6×10^{-3} cm²)

levels and hence cause a shift of the recombination spectrum to higher energies.

The photon energy of the peak of the spontaneous emission is plotted as a function of temperature in Fig. 3.14 and compared with the band gap of pure gallium arsenide as determined by Sturge (1962). The peak emission observed at low current densities accurately follows the band gap energy but is displaced to a lower energy. The magnitude of this displacement depends on the doping levels in the diode and in this case amounts to 0·07 eV. The exact nature of the energy levels involved in the radiative transition has been considered in some detail in Chapter 2; for the present purpose it is sufficient to note that the wavelength of the emission is

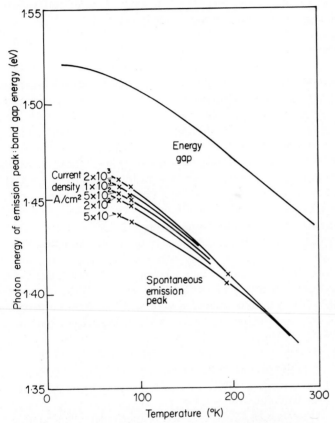

Fig. 3.14. The variation of spontaneous emission peak
with temperature and current density

consistent with a model in which recombination takes place between valence and conduction bands, which are considerably perturbed by band tails produced by the high concentration of donors and acceptors present. These tails extend into the band gap and give rise to radiative recombination with a photon energy somewhat less than the band gap of the pure material.

If the spontaneous emission from a laser diode is examined with sufficient resolution, the formation of spectral modes can be observed (Fig. 3.15a). It is believed that the formation of these modes is an inter-ference effect due to the reflection of radiation at the faces at the cavity. Vilner (1962) predicted that such an effect would be observed when radiation is spontaneously emitted by a medium contained within an optical cavity (Nathan and coworkers, 1963).

3.4.2 Laser emission

Radiation which is fed back into the active region of the diode will produce stimulated transitions. At sufficiently high current densities these stimulated transitions will produce a gain which is sufficient to overcome the losses in the system and coherent oscillation will be observed. This is the threshold condition which was discussed in Section 3.1 and at this point the energy in one or more of the spectral modes, near the peak of spontaneous emission, will increase rapidly (Fig. 3.15b). At currents slightly above threshold the spectrum will be dominated by the stimulated emission. Careful observation shows that the threshold current defined in this way agrees closely with that determined by visual observation or from a plot of power output as a function of current.

At currents well above threshold the laser spectrum is complex. Fig. 3.16 shows a typical example in which emission occurs simultaneously in several modes. As the current through the diode is increased the complexity of the spectrum increases and more modes appear. Although the total power emitted from the laser increases with current it does not necessarily follow that the power in any particular mode will increase monotonically (Fig. 3.17). At the present time it is not possible to predict the relative amplitudes of the observed modes although it is suggested that this is influenced by inhomogeneities within the device.

Provided that thermal effects are avoided the wavelengths of the spectral modes remain constant as the junction current is increased. The separation of these modes, $\Delta\lambda$, is inversely proportional to the length of the laser cavity and it is found that, at 77°K, $l\Delta\lambda \approx 0.07$ cm Å (Kingsley and Fenner, 1963; Burns and coworkers, 1963). A typical diode with a cavity length of 0.05 cm will have a mode separation of 1.6 Å approximately and this can easily be resolved on a good grating spectrometer.

If it is assumed that the modes which are observed are simple longitudinal modes of the laser cavity, their wavelength will be given by

$$n\lambda = 2L\mu \qquad (3.13)$$

where n is an integer and μ is the refractive index of the cavity material.

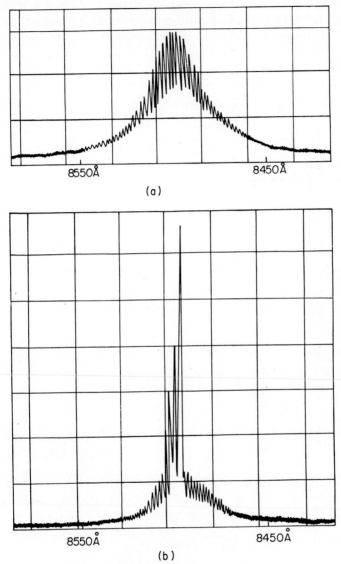

(a)

(b)

Fig. 3.15. Laser spectra near threshold
(a) Below threshold
(b) Above threshold

The wavelength spacing between adjacent modes can be obtained by subtracting values of λ corresponding to successive values of n. However, since n is large, this can be obtained by differentiating (3.13) with respect to n and putting $\Delta n = 1$, bearing in mind that the refractive index of the material is a function of wavelength so that dispersion must be taken into account.

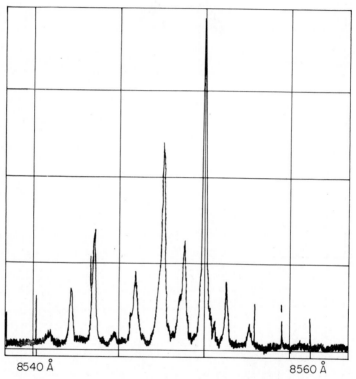

8540 Å   8560 Å

FIG. 3.16. Laser spectrum well above threshold $(T_0 = 77° \text{ K})$

Hence we have

$$\Delta\lambda = \frac{\lambda^2}{2\mu L\left(1 - \frac{\lambda}{\mu}\frac{\partial\mu}{\partial\lambda}\right)} \tag{3.14}$$

Using the refractive index data given by Marple (1964) good agreement is obtained between the observed mode spacing and that calculated from (3.14).

Since the optical properties of a semiconductor are temperature dependent one would expect the wavelengths of the laser modes to depend on the temperature at which the device is operated. This is found to be the

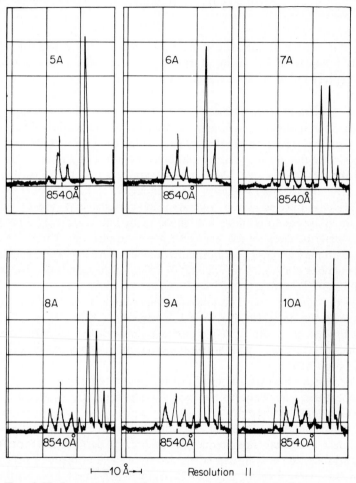

FIG. 3.17. Laser spectra taken at successively increased currents. (Junction area $1·6 \times 10^{-3}$ cm², temperature 77° K)

case and experiments show that the wavelength has a temperature coefficient of approximately 0·04 Å/deg. at 77°K. A value for this temperature coefficient can be calculated from (3.13) by differentiating with respect to T, bearing in mind that the refractive index is a function of both temperature and wavelength.

Hence we obtain

$$\frac{1}{\lambda}\frac{d\lambda}{dT} = \frac{\dfrac{1}{\mu}\dfrac{\partial \mu}{\partial T} + \dfrac{1}{L}\dfrac{dL}{dT}}{1 - \dfrac{\lambda}{\mu}\dfrac{\partial \mu}{\partial \lambda}} \qquad (3.15)$$

In this expression the term representing the thermal expansion of the cavity is negligible and, using Marple's refractive index data, one obtains good agreement with experimental results.

3.4.3 Spectra at high resolution

When the spectrum of a laser is examined with a resolution of the order of 0·01 Å it is frequently found that each of the simple longitudinal modes discussed in the previous paragraphs consists in fact of a family of modes (Sorokin and coworkers, 1963) and this can be attributed to oscillations occurring in non-axial modes in the laser cavity.

Carefully selected lasers operated at currents not greatly above threshold appear to oscillate in a single mode and several attempts have been made to measure the linewidth of such radiation. Results of such linewidth measurements are normally quoted in terms of a frequency and it should be noted that a wavelength of 8500 Å corresponds to a frequency of $3·5 \times 10^{14}$ Hz and a linewidth of 1 GHz to 0·03 Å.

Using a simple Fabry–Perot interferometer Engeler and Garfinkel (1964) estimated the linewidth of a laser to be less than 300 MHz at 77°K and Armstrong and Smith (1964) reported a linewidth of 50 MHz from a device operating at 15°K although the single mode power in the latter case was only 1 mW.

To measure narrower linewidths heterodyne techniques can be used. The beat frequency between two cavity modes separated by 2 Å would be 100 GHz and the linewidth of this beat frequency would be a measure of the mode linewidth. However, no optical detectors are available which operate at sufficiently high frequencies. In order to perform this experiment an anti-reflection coating can be applied to one end of the laser which can then be operated in an external cavity giving a lower beat frequency. Using such a technique Crowe and Craige (1964) operated a diode in a 7·2 cm cavity giving a heat frequency of 1·5 GHz and obtained a linewidth of 10 MHz.

Linewidths of lasers have also been measured using a Michelson interferometer with a path difference of 3000 ft and heterodyne detection (Ahearn and Crowe, 1966). With this technique linewidths as low as 150 KHz were obtained.

3.4.4 Temperature and impurity effects

Since laser emission occurs at a photon energy near the peak of the spontaneous emission, the wavelength at which laser emission occurs will show the same temperature dependence as the spontaneous emission. The

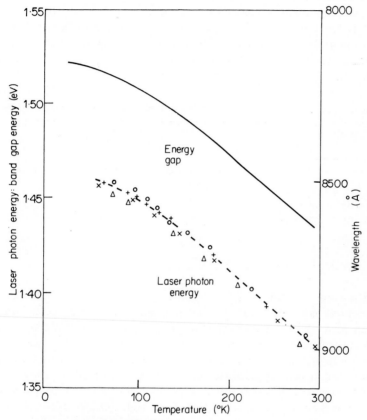

Fig. 3.18. The photon energy of laser emission as a function of temperature. The typical scatter of results for several diodes is shown

wavelength of the laser emission is thus approximately 8500 Å at 77°K and 9000 Å at 300°K.

The variation of photon energy with temperature for several lasers is shown in Fig. 3.18 for the temperature range 77 to 300°K and it can be seen that, for these lasers, emission occurs at a photon energy 0·07 eV less than the band gap energy over the whole temperature range.

The photon energy of the laser emission also depends somewhat on the nature and concentration of the dopants used in the diode construction (Dousmanis and coworkers, 1963) and these effects can be understood in terms of the band tail model of a laser discussed by Stern (1966). According to this model radiative transitions occur between states which form 'tails' on the valence and conduction bands and these tails are produced by the high concentration of donors and acceptors present in the material. These tails can be described by an energy parameter which gives a measure of the depth to which the tail extends in the band gap. Stern has calculated the band tail energies at 80°K given in Table 3.4.

TABLE 3.4. Valence and conduction band tail energies calculated by Stern (1966).

Impurity concentration (cm^{-3})		Band tail energy (meV)		
Donor	Acceptor	Conduction band	Valence band	Total
3×10^{17}	$3 \cdot 3 \times 10^{18}$	10·1	33·4	43·5
1×10^{18}	4×10^{18}	18·7	37·4	56·1
3×10^{18}	6×10^{18}	33·4	47·3	80·7
1×10^{19}	$1 \cdot 3 \times 10^{19}$	65·6	74·8	140·4
3×10^{19}	$3 \cdot 3 \times 10^{19}$	125	131	256

The diodes whose characteristics are shown in Fig. 3.18 had a donor concentration 2×10^{18} cm^{-3} so that the band tail energies calculated by Stern correlate resaonably with the amount that the laser photon energy is less than the band gap energy of pure gallium arsenide. However, this is an oversimplification since, to predict the energy at which laser action will occur, one must calculate the variation of gain in the system with energy and hence determine the energy at which maximum gain, and laser action, will occur. These calculations have been considered in some detail in Chapter 2.

The figures in Table 3.4 indicate the variation of laser photon energy or wavelength which is to be expected if the impurity concentrations in the diode are varied. For example, a variation of donor concentration from 1×10^{18} to 2×10^{18} cm^{-3} would cause an energy change of 25 meV corresponding to a wavelength change of 180 Å. Since, with present technology it is not possible to control the doping level in gallium arsenide to better than 10%, appreciable variations in the wavelengths of laser emission in different diodes is to be expected and in practice a wavelength spread of ± 25 Å is observed in a batch of nominally identical devices.

It has been shown that the threshold current density of a laser depends on its cavity length. A shorter cavity will have a higher threshold current density and this implies that a higher density of minority carriers must be injected into the system to produce sufficient gain for laser action to occur. One would thus expect this increasing minority carrier concentration to populate the conduction band to higher energies and thus cause a shift of spectrum to higher energies. This effect is small and will normally be masked by random variations in laser properties. However, in experiments with a series of carefully constructed diodes of different lengths Yonezo and coworkers (1968) have shown that the effect can be observed.

3.4.5 Polarization

The emission from a semiconductor shows in general polarization effects. The electromagnetic theory of a semiconductor laser predicts that the structure will support both TE and TM electromagnetic modes and that the threshold for these modes will be similar. Hence one would expect to observe radiation polarized either parallel or perpendicular to the junction plane.

In practice the situation is somewhat more complicated than this. In the simplest case, radiation in which the **E** vector is perpendicular to the junction plane is observed but polarization in a direction perpendicular to this is also observed. However, elliptical polarization also occurs and frequently different regions of a laser will emit radiation of different polarizations.

In a detailed study of the radiation from laser diodes Alyanovskii and coworkers (1966) found no correlation of the polarization with any of the junction properties which could be controlled and they concluded that the apparently random polarization was controlled in some way by defects or inhomogeneities within the laser.

The polarization of laser radiation is a potentially useful property which could be used in some applications to separate low intensity laser radiation from any background radiation. However, at the present time it has not been possible to exploit this potential due to the unpredictable polarization of emission from laser diodes.

3.4.6 Harmonic generation

Harmonic generation in gallium arsenide lasers has been observed. The second harmonic occurs in the blue region of the spectrum and is strongly absorbed in gallium arsenide; only radiation produced near the surface of the diode is emitted and the power observed is small.

Armstrong and coworkers (1963) detected a second harmonic power of $1 \cdot 6 \times 10^{-13}$ W with a fundamental power of 5×10^{-3} W and found that the harmonic power varied as the square of the fundamental power. Garfinkel and Engeler (1963) made a detailed study of the spectrum of the second harmonic radiation. The fundamental radiation will have cavity frequencies $f_1, f_2, f_3, \ldots$ which are equally spaced. The second harmonic will have frequencies $2f_1$, $2f_2$, f_1+f_2, etc., and several of these harmonics (e.g. $2f_3$, f_2+f_4, f_1+f_5) will have the same frequencies. In this way 13 second harmonic frequencies were correlated with 11 fundamental frequencies. Malstrom and coworkers (1964) obtained 10^{-5} W of second harmonic power with a fundamental power of 20 W and estimated a non-linear susceptibility for gallium arsenide of $2 \cdot 7 \times 10^{-6}$ e.s.u. which is considerably greater than that of KDP.

3.5 Continuous operation

3.5.1 Conditions for C.W. operation

Gallium arsenide lasers are most frequently operated under pulsed conditions and in many instances the power densities at which devices are required to operate make continuous operation impossible. However, to obtain the maximum stability of power output and the narrowest spectral linewidths continuous operation at a carefully controlled temperature is essential and in this section we will discuss the conditions for the continuous operation of a device.

If a continuous current, I, is passed through a laser the temperature at the junction will rise to an equilibrium value, ΔT, above the ambient temperature, T_0. If the threshold current of the device at T_0 is I_0 then the threshold current at temperature $T_0 + \Delta T$ can be expressed as $I_0' = I_0 f(\Delta T)$ and the device will only be in a lasing state if $I_0' < I$. This condition can be illustrated on a graph showing temperature versus current (Fig. 3.19). On such a plot two curves can be drawn for any device:

[i] *Junction temperature versus current.* If we ignore the power which is radiated, the power dissipated in a device is given by

$$P = IV + I^2 R \tag{3.16}$$

where V is the junction voltage and R is the series resistance of the device.
The junction temperature rise is given by

$$\Delta T = \Theta P \tag{3.17}$$

where Θ is the thermal impedance of the device.

[ii] *Threshold current versus temperature.* For most practical purposes the threshold current can be taken to be a cubic function of the junction temperature so that

$$I' = I_0(1 + \Delta T/T_0)^3 \qquad (3.18)$$

Continuous operation of the device is possible only if these two curves intersect and then only over a limited range of currents as shown in

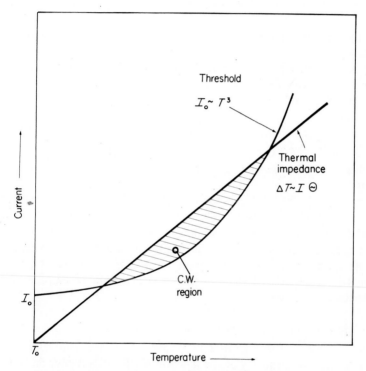

Fig. 3.19. Threshold current versus temperature and junction temperature versus current for a laser, showing region of continuous operation

Fig. 3.19. It is of interest to estimate the requirements of a laser for continuous operation and in doing so we will assume that one of the two terms in (3.16) is dominant and it will be shown that this corresponds to two situations of practical interest.

[1] At low temperatures devices operate at relatively low currents so that the Joule heating in a device can be neglected. In this case the

temperature rise at the junction is given by $\varDelta T = \Theta IV$ and continuous operation of the device is possible if

$$I_0\left(1+\frac{\varDelta T}{T_0}\right)^3 \leqslant \frac{\varDelta T}{\Theta V} \tag{3.19}$$

For this to be possible we must have

$$\frac{I_0 V \Theta}{T_0} \leqslant 0\cdot 15 \tag{3.20}$$

and it is thus convenient to define a figure of merit $q = I_0 V \Theta / T_0$ for a device.

[2] At higher temperatures Joule heating will be the predominant dissipation in the device and the junction temperature rise will be $\varDelta T = \Theta I^2 R$. In this case continuous operation is possible if

$$I_0\left(1+\frac{\varDelta T}{T_0}\right)^3 \leqslant \left(\frac{\varDelta T}{\Theta R}\right)^{\frac{1}{2}} \tag{3.21}$$

For this to be possible we must have

$$\frac{I_0^2 R \Theta}{T_0} \leqslant 0\cdot 067 \tag{3.22}$$

and it is convenient to define a second figure of merit $s = I_0^2 R \Theta / T_0$.

We thus have two conditions which must be satisfied if a device is to operate continuously. These conditions involve the series resistance and thermal impedance of devices so that it is necessary to consider these properties in more detail.

3.5.2   *The measurement of device thermal properties*

Several techniques can be used to estimate the thermal impedance of a laser. All require estimates of the junction temperature under conditions of known power dissipation and this temperature can be inferred either from observations of the wavelength of emission or from observations of the device threshold. We will not assume that the device can be operated continuously so that the techniques described must be applicable to devices operating in a pulsed mode.

[1] Suppose a direct current is passed through a device and produces a dissipation IV. If a current pulse is then superimposed on this direct current the current pulse can be of sufficient amplitude to bring the device into the lasing state. The wavelength of emission can be measured and hence the junction temperature estimated.

[2] In the above experiment one can observe the amplitude of the current pulse, I', required to bring the device into the lasing state. If the threshold current is a known function of temperature then the junction temperature can be estimated and the device thermal impedance calculated.

[3] If a device is operated with a pulse current I and a pulse length τ at a frequency f the junction temperature rise will be

$$\Delta T = (IV + I^2 R)\tau f \Theta \tag{3.23}$$

which can be estimated from observations of the emission wavelength. Hence values of $(1 + IR/V)\Theta$ can be obtained for various values of I and both R and Θ can be determined.

Some results of measurements using technique [3] are shown in Fig. 3.20. The thermal impedance of this device is 32 deg.K/W. The electrical resistance deduced from these measurements ($0.07\ \Omega$) agrees well with the measured series resistance ($0.09\ \Omega$).

It has been assumed that the same value of thermal impedance is to be applied to both the junction dissipation and the Joule dissipation. This is not strictly true and one should write

$$\Delta T = IV\Theta_1 + I^2 R\Theta_2$$

or

$$\Delta T = (IV + I^2 R\Theta_2/\Theta_1)\,\Theta_1 \tag{3.24}$$

and the assumption of a unique value for the thermal impedance of a device is acceptable if one uses an effective series resistance $R' = R\Theta_2/\Theta_1$ in thermal calculations. In practical cases R' will be smaller than R. For instance, in the above example the measured device series resistance was $0.09\ \Omega$ but the error involved in neglecting the difference between this and the effective (thermal) value of $0.07\ \Omega$ would normally be negligible in this type of calculation.

3.5.3 The calculation of device thermal properties

[i] *Thermal impedance.* A simple laser structure would consist of a laser dice mounted on a massive heat-sink of high thermal conductivity material. In such a simple structure the thermal impedance between the junction and the heat-sink can be considered as two components in series.

[1] The thermal impedance of the gallium arsenide between the junction and the heat-sink. This is given by

$$\Theta_A = \frac{t}{K'wL} \tag{3.25}$$

where t is the thickness of the gallium arsenide, w, L are the dimensions of the junction and K' is the thermal conductivity of GaAs.

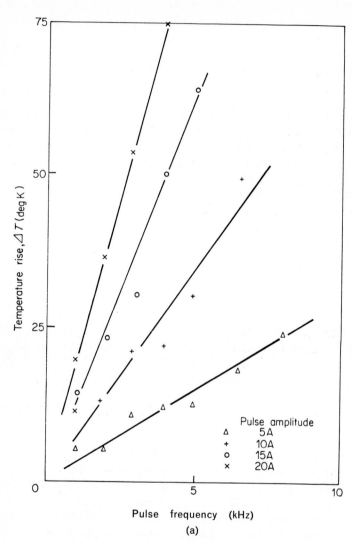

FIG. 3.20(a). Determination of thermal impedance using Eqn. (3.23). The slope of these curves gives values of $\Theta(1+IR/V)\tau$ for corresponding values of I. (pulse length 5μ sec; pulse amplitude 5, 10, 15, 20 A)

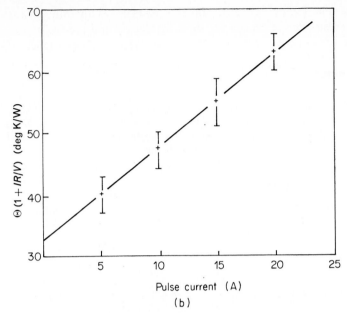

FIG. 3.20(b). Determination of thermal impedance using Eqn. (3.23) and values of $\Theta(1+IR/V)$ from Fig. 3.20 (a) giving $\Theta = 32$ deg. K/W and $R = 0{\cdot}07\ \Omega$

[2] The thermal spreading resistance into the heat-sink. For a rectangular contact this is given by (Torrey and Whitmer, 1948)

$$\Theta_B = \frac{\ln(4L/W)}{2\sqrt{\pi K''L}} \qquad (3.26)$$

where K'' is the thermal conductivity of the heat-sink and a rectangular contact has been approximated to an ellipse of equal area and aspect ratio.

The relative magnitude of the components can best be illustrated by some typical examples of lasers mounted on copper heat-sinks (Table 3.5).

From the above formulae it can be seen that to obtain a device with a low thermal impedance it should be made as thin as possible and that, for a given junction area, the thermal spreading resistance will be reduced if the aspect ratio is as large as possible. The thermal properties of a device will be further improved if it is mounted between two heat-sinks to cool the junction from both sides.

[ii] *Series resistance.* The electrical resistance of a device will be due to the resistivity of the bulk material ($\sim 0{\cdot}001\ \Omega\,\text{cm}$) and to the resistance

5

between the gallium arsenide and the contacts. With thin devices the contact resistance is usually the more significant and the resistance of a contact of unit area will be $\sim 10^{-4} \, \Omega \text{cm}^2$. Hence the resistance of the devices considered in the previous paragraph will be 0·01 and 0·2 Ω, respectively.

TABLE 3.5. The estimated thermal impedance of three representative lasers (A, B and C) showing the contributions due to: (a) the gallium arsenide; (b) the thermal spreading resistance into a copper heat-sink and (c) the total thermal impedance of the structures.

Example		A	B	C	
Laser	L	0·1	0·05	0·05	cm
dimensions	w	0·1	0·01	0·01	cm
	t	0·0025	0·0025	0·0025	cm
Operating temperature	T	77	77	300	°K
Thermal	GaAs	2·5	2·5	0·5	W/cm deg.K
conductivity	Cu	6·6	6·6	4·0	W/cm deg.K
Thermal	(a) GaAs	0·1	2	10	deg.K/W
impedance	(b) Cu	0·7	3	5	deg.K/W
	(c) Total	0·8	5	15	deg.K/W

[iii] *Figures of merit.* If we assume threshold current densities of $10^3 \, \text{A/cm}^2$ and $4 \times 10^4 \, \text{A/cm}^2$ at 77 and 300°K, respectively, we can now estimate the figures of merit applicable to our previous examples.

TABLE 3.6. The estimated thermal figures of merit of three representative lasers.

Example		A	B	C	
Temperature		77	77	300	°K
Thermal impedance		0·8	5	15	deg.K/W
Series resistance		0·01	0·2	0·2	Ω
Threshold current		10	0·5	20	A
Figure of merit	q	0·16	0·05	1·5	
	s	0·01	0·003	4	

Comparing the values of q and s with the requirements for continuous operation it can be seen that continuous operation at 77°K will be feasible for the smaller device but that considerable improvements are required in devices if they are to be operated continuously at room temperature.

Although these estimates are crude and optimistic, it will be seen in the next section that the conclusions are in reasonable agreement with the present state of the art of laser development.

3.5.4 C.W. power output

It is, of course, important not only that a device should be able to operate in a continuous mode, but that it should also produce as high a power output as possible.

The power output from a device can be expressed as

$$P = (I - I_0') V \Delta \eta \qquad (3.9)$$

where I is the continuous current through the device and I_0' is the threshold current in the operating condition.

The temperature at the junction, and hence the operating threshold current, can be calculated from the equations given in Section 3.5.1 and it is thus possible to predict the power output of a device as a function of drive current.

Pilkuhn and Guettler (1968) have considered the design of a device to give a maximum power output using ideas similar to those discussed in the previous sections but assuming that the power input to a device was limited by the maximum junction temperature which could be tolerated. They take into account the variation of laser threshold and efficiency with cavity length and reflectivity. These calculations predict an optimum value for the cavity length and width and show that there will be an optimum value for the reflectivity.

However, although design studies such as these form a very useful guide to the construction of laser diodes, in practice the performance of devices is limited by imperfect electrical and thermal contacts and other technical problems. In these circumstances the design of devices usually proceeds on a semiempirical basis.

3.5.5 Experimental devices

The earliest gallium arsenide lasers were made with rather crude alloyed contacts which were far from ideal and introduced a considerable thermal impedance to the structure. However, even with such devices continuous operation was possible with lasers cooled to liquid helium or hydrogen temperatures.

Continuous operation at 27°K was reported by Dousmanis and co-workers (1963) and with careful design and construction considerable power can be obtained at these low temperatures. Engeler and Garfinkel (1964) obtained a continuous power output of 3 W from a device operating at 20°K and 7 W has been obtained from a device cooled to 4·2°K (Clifton and Debye, 1965).

Continuous operation at liquid nitrogen temperatures and above is somewhat more difficult to achieve although the problem has been eased

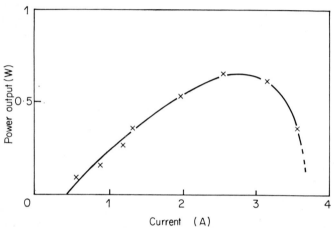

FIG. 3.21. Power output versus current for a continuously operating laser ($T_0 = 77°$ K)

considerably by the contacting techniques described by Marinace (1966) (see Chapter 5). Using these techniques continuous operation of small devices at 77°K is readily obtainable and output powers of 1 W or more can be obtained.

The power output characteristics of a typical device operating continuously at 77°K are shown in Fig. 3.21. In this figure the power output is plotted as a function of the direct current through the device and it can be seen that an appreciable output is obtained only over a limited range of currents corresponding to the device being in a lasing state.

Considerable efforts have been made to extend the temperature range over which continuous operation of a device is possible and Dyment and D'Asaro (1967) have reported C.W. operation at a temperature of 200°K. This was achieved by mounting a very thin and narrow device on a diamond heat-sink, but unfortunately the techniques necessary to make such devices appear to result in devices with poor efficiency.

Continuous operation of devices at room temperature has still not been achieved and appreciable powers from C.W. devices at room temperature will only be achieved if all the properties of devices can be considerably improved. For example, the device C in the previous section would operate continuously if its threshold current density were reduced by a factor of five and both its electrical resistance and thermal impedance by a factor of two. Such improvements are still some way off.

3.6 Pulsed operation

In most practical applications gallium arsenide lasers are used in a pulsed mode of operation. In a previous section we have considered the properties of lasers under somewhat idealized conditions and assumed that the temperatures of the laser were carefully specified. In this section we will see that pulsed operation produces several thermal effects which will be considered in some detail and it will be convenient to consider these effects in two categories.

If the properties of a laser are examined during a single pulse, transient effects can be observed and these can be correlated with the temperature rise in the laser during the pulse. Alternatively, if the laser is operated with a continuous series of short pulses quasi-equilibrium conditions are obtained and it will be shown that these conditions can be described using the concept of thermal impedance which was discussed in the previous section.

3.6.1 Quasi-equilibrium conditions

Consider a laser which is operated with current pulses of amplitude I, pulse length τ and frequency f. It will be assumed that the pulse length is short and we will return later to a consideration of the pulse lengths required to meet this criterion.

Following the approach used in the previous section, the power dissipated in the laser is given by

$$P = (IV + I^2 R)\tau f \qquad (3.27)$$
$$= (IV + I^2 R)\delta$$

where $\delta = \tau f$ is the duty cycle at which the device is operating. The device junction temperature, threshold current and peak power output under these conditions are given, respectively, by

$$T = T_0 + \Delta T = (IV + I^2 R)\delta\Theta \qquad (3.28)$$
$$I_0' = I_0(1 + \Delta T/T_0)^3 \qquad (3.29)$$
$$P = (I - I_0')V\Delta\eta \qquad (3.30)$$

The mean power $\bar{P}$ emitted by the device is given by the product of the peak power and the duty cycle:

$$\bar{P} = P\delta = (I - I_0')\, V\, \varDelta\eta\, \delta \tag{3.31}$$

The performance of a typical device calculated in this way is shown in Fig. 3.22. Although the model which has been used is relatively crude the

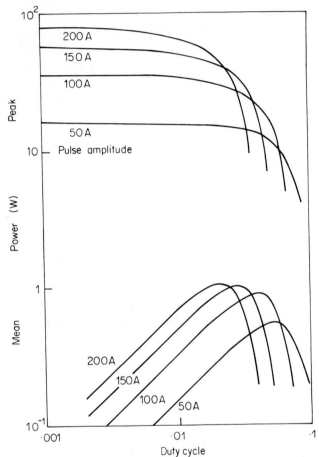

FIG. 3.22. Performance of a laser under pulsed conditions. Calculated from Eqns. (3.30) and (3.31) with $I_0 = 10\,\text{A}$, $\Theta = 5\,\text{deg.K/W}$, $R = 0{\cdot}01\,\Omega$, $T_0 = 77°\text{K}$

predicted results agree well with those found experimentally and it can be seen that there is considerable scope for the optimization of device operating parameters to meet particular requirements.

Lasers operated under pulsed conditions can of course give considerably higher peak powers than devices operating continuously. Lasers operated at 77°K or below can produce a peak power output of 100 W and at this peak power devices can be operated at a duty cycle such that a mean power of 1 W or more is obtained.

At room temperature peak powers of greater than 50 W have been reported but only with pulse lengths ~0·1 μsec. Most of the devices used at room temperature have been made by liquid epitaxial techniques (Nelson and coworkers, 1964; Nelson, 1967), but Carlson (1967) has shown that diodes made by diffusion processes can have threshold currents less than 4×10^4 A/cm² and a high quantum efficiency.

The duty cycle at which devices can operate at room temperature is limited to about 0·1% and the mean power is correspondingly low. By making use of cooling by water or forced convection Dousmanis and Gross (1966) have operated a device at a pulse repetition rate of 50 kHz and hence obtained a mean power of 50 mW.

3.6.2 Transient effects

While a current pulse is applied to a laser diode the temperature at the junction will rise. This temperature rise is manifested as a decrease in the power emitted from the device and complex variations in the spectrum of the emitted radiation (Gooch, 1965; Konnerth, 1965a; Gonda, 1965).

The transient spectral changes are due to the temperature dependence of two properties of the diode: (*a*) the temperature dependence of band gap and spontaneous emission spectrum which causes the laser emission to shift with a temperature coefficient of approximately 2 Å/deg.K; (*b*) the temperature dependence of the refractive index which causes each individual mode to shift with a temperature coefficient 0·5 Å/deg.K.

Both of these effects can be seen in Fig. 3.23 which shows the spectral modes which were observed at various times during the application of a long (100 μsec) current pulse to a laser.

A detailed analysis of the rate at which the temperature of a diode rises during an operating pulse is a complex problem which depends on the structure of the particular diode although certain general conclusions can be reached. Heat is generated both at the junction and, due to the resistivity of the gallium arsenide, in the bulk of the device. For very short pulse lengths heat will not have time to diffuse away from the region in which it is produced. Under these adiabatic conditions the temperature rise will be proportional to the pulse length. During somewhat longer pulses thermal diffusion will occur and the junction temperature will increase as the square root of the pulse length. Ultimately thermal equilibrium would be reached although this would normally not occur until the temperature had risen so high that laser action was no longer possible.

In order to estimate the initial rate of rise of temperature in a device one must know the thickness of the region in which the junction power is dissipated. Evidence suggests that this region is $\sim 50\ \mu$ wide at 77°K and

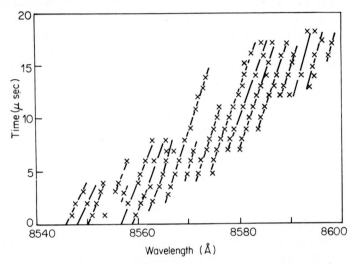

FIG. 3.23. Transient spectra showing principal modes observed during a 10 A, 20 μsec pulse. (Diode area $1 \cdot 6 \times 10^{-3}$ cm², $T_0 = 77°$ K)

this is considerably wider than the depletion width of the junction due to the removal of energy away from the immediate vicinity of the junction by photon transport processes. The value quoted is likely to be considerably less at 300°K since the radiative processes are less efficient, and the absorption coefficients higher, at this temperature.

3.6.3   Pulse requirements

In order to achieve laser action in a device which will not operate continuously, a current pulse of sufficient amplitude and with a sufficiently fast rise-time must be applied to the device. It is possible to make an estimate of rise-time required in the following way.

At a current density, j, the dissipated power density in the junction region is given by

$$\frac{jV}{D} + \frac{j^2}{\sigma} \tag{3.32}$$

where D is the thickness of the region in which the junction dissipation occurs and σ is the electrical conductivity of the bulk gallium arsenide. Even under the most extreme conditions, with current densities of 10^5 A/cm^2 the first term of the above expression is dominant and hence, in the adiabatic condition, the rate of the temperature rise at the junction is

$$\frac{dT}{dt} = \frac{jV}{Dc\rho} \tag{3.33}$$

where c is the specific heat of the gallium arsenide and ρ is the density of gallium arsenide.

If a triangular current pulse $j = \phi t$ is applied to the laser the temperature at a time t' will be given by $T_0 + \Delta T$ with

$$\Delta T = \int_0^{t'} \frac{Vj\,dt}{Dc\rho} = \frac{VJ^2}{2\phi Dc\rho} \tag{3.34}$$

where J is the current density at t'.

At this instant the device threshold current density is

$$J_0' = J_0(1 + \Delta T/T_0)^3 \tag{3.35}$$

and if the device is just to be in a lasing state

$$J_0' = J$$

or

$$J_0\left(1 + \frac{\Delta T}{T_0}\right)^3 = \left(\frac{2\phi Dc\rho\,\Delta T}{V}\right)^{\frac{1}{2}} \tag{3.36}$$

If this is to be satisfied

$$\frac{J_0^2 V}{2\phi Dc\rho T_0} = 0 \cdot 067 \tag{3.37}$$

and

$$J = 1 \cdot 7 J_0 \tag{3.38}$$

Putting

$$\phi = J/t' \tag{3.39}$$

we have

$$t' = \frac{2 \times 0 \cdot 067 c\rho DT_0}{J_0 V} \tag{3.40}$$

If the device is just to reach threshold the required pulse rise-time is given by (3.40) and the pulse amplitude by (3.38).

At room temperature, taking $J_0 = 5 \times 10^4$ A/cm^2 and $D = 10\ \mu$, we have $t' = 0 \cdot 8\ \mu$sec while at 77°K, with $J_0 = 10^3$ A/cm^2 and $D = 50\ \mu$, we have $t' = 30\ \mu$sec (this latter figure is probably beyond the limits in which the adiabatic model is valid).

It must be emphasized that the pulse rise-times which have been estimated are those which would enable the device to just reach the threshold condition. If the device is to be used with any efficiency and with a useful power output rise-times at least an order of magnitude faster than these will be required and, in practice, pulse lengths of $10\ \mu$sec and $0 \cdot 1\ \mu$sec are suitable at 77 and $300°$K, respectively.

Broom (1968) has extended this type of calculation to include thermal diffusion away from the laser junction. He has shown that it is possible to predict the output power pulse shape and obtains good agreement with experimentally observed pulse shapes.

3.6.4   Damage and degradation

When lasers are operated at high peak powers damage to the polished cavity faces can occur. This damage appears as a line of pits or fractures along the line of the junction (Cooper and coworkers, 1966).

At $77°$K this damage occurs when the power emitted by the device exceeds 500 W/cm of junction width, which corresponds to a local power density of 10^6 W/cm^2 transmitted through the cavity face. The power density required to produce such damage varies somewhat with the structure of the device, but below this figure it appears that devices operate reliably with no reported degradation in performance. At room temperature such damage can undoubtedly occur although no systematic studies of it have been reported.

Kressel (1968) has reported another form of degradation occurring in lasers operated at room temperature. Lasers operated at high current densities show a gradual decrease in output with no visual damage to the cavity face. This degradation is accompanied by a 'softening' of the diode characteristics and appears to be an effect occurring over the whole area of the junction. Experiments suggest that this degradation is due to the high current densities used rather than high optical flux densities and it can be avoided if the diodes are operated at currents not exceeding two to three times the threshold value.

Further work by Kressel and Byer (1969) on diodes operated at room temperature has shown that during operation the threshold current of a device increases and the device efficiency decreases. Diodes with a non-uniform near-field emission pattern are more prone to degradation than uniform devices and the best devices can operate at pulse current densities of 10^5 A/cm^2 with no appreciable degradation after several hundred hours operation with 200 nsec pulses at a repetition rate of 10 kHz. On the other hand, the output of poor devices can halve after similar operation for only a few hours.

3.7   Pressure effects

3.7.1   *Hydrostatic pressure*

The effect of hydrostatic pressure on the spectrum of a laser is similar to the effect of temperature (Feinleib and coworkers, 1963; Fenner, 1963; Stevenson and coworkers, 1963). The spontaneous emission has a pressure coefficient of 10^{-5} eV/atm. With these values for the pressure coefficients a pressure change of 50 atm produces wavelength changes comparable to those produced by a temperature change of $1°K$, so that in experiments on pressure effects careful temperature control is necessary.

The pressure dependence of the spontaneous emission is directly related to the pressure dependence of the band gap of the semiconductor and the pressure dependence of a lasing mode can be related to the variation of refractive index with pressure. No direct measurements of the pressure dependence of refractive index are available, but Fenner has shown that this can be estimated from the refractive index data given by Marple (1964) if it is assumed that the variation of refractive index with both temperature and pressure is due to the movement of the band gap with temperature or pressure. This is equivalent to assuming that the dispersion curves for different temperatures or pressures can be obtained by translating the dispersion curve along the energy axis by the change in band gap produced by the temperature or pressure change. In this way Fenner has shown that the observations on mode spectra are consistent with the available data on the band gap and refractive index of gallium arsenide.

3.7.2   *Uniaxial pressure*

The application of a uniaxial stress perpendicular to the junction plane of a laser diode has been found to produce a shift in the spontaneous emission wavelength and a lowering of the laser threshold (Ryan and Miller, 1963). The results depend on the orientation of the diode and can be interpreted by considering the departure from cubic symmetry which results from uniaxial strain (Miller and coworkers, 1964).

3.8   Modulation

3.8.1   *Pulse modulation*

The response of a laser to a current pulse is of considerable interest in many applications. For example, in an optical radar system the accuracy with which a distance can be measured is limited by the rise-time of the light pulse used. To measure a distance to an accuracy of 10 cm a rise-time of the order of 10^{-9} sec must be used.

Konnerth and Lanza (1964) have shown that there is a finite delay between the application of a current pulse to a laser and the emission of stimulated radiation. This is due to the finite time required to establish an inverted population and decreases with the amplitude of the applied current pulse. The delay, t, follows the equation

$$t = \tau \ln \left(\frac{I}{I-I_0} \right)$$

where τ is the spontaneous life time, I_0 is the threshold current and I is the amplitude of the applied pulse.

At 77°K τ is ~2 nsec and the rise-time of the light pulse once laser action is established is <0·2 nsec.

At room temperature the delays observed are considerably longer than those at 77°K and values of several hundred nano-seconds have been reported (Winogradoff and Kessler, 1964; Konnerth, 1965b).

Dyment and Ripper (1968) have shown that in some cases the increase in delay from a low value at 77°K to a high value at room temperature occurs at a clearly defined temperature and at this temperature the threshold current versus temperature curve shows a discontinuity.

The delays which are observed could be explained by the presence of traps in the laser material. These traps would have to be filled by injected electrons before an inverted population could be established. However, the density of traps which must be postulated to account for the magnitude of the observed delays is unreasonably large.

Fenner (1967) has proposed an alternative model in which laser action is delayed, not by traps which must be filled before inversion is possible but by optical absorption centres which must be saturated before oscillation can occur. However, at the present time there are objections to both these models.

It has been found that the delays which are observed at room temperature vary widely according to the techniques by which the devices are made. Diffused diodes often show long delays which can be reduced by suitable annealing techniques. In diodes made by epitaxial techniques the delays are usually small but no convincing reason for this difference has been suggested.

3.8.2 Amplitude modulation

Once a laser is above threshold it will respond rapidly to variations in the amplitude of drive current and it is probable that at the present time the highest modulation frequencies observed have been limited by the experimental techniques available.

A laser operating continuously at 4·2°K has been modulated at 11 GHz (Goldstein and Weigand, 1965) but attempts to modulate lasers at higher frequencies have shown some interesting resonance effects (Takamiya and coworkers, 1968). If the optical resonant (mode) frequency of a laser is v, then modulation at a frequency f will produce side-band frequencies $v+f$. It was found that efficient amplitude modulation at the frequency f was only possible if the side band produced, $v+f$, was also a resonant frequency of the laser. In the experiments quoted the separation of laser modes was 1·1 Å and hence modulation at 46 GHz could be obtained. It would thus appear that high frequency modulation of a laser will require a critical cavity length to give the required mode separation and optical resonance at the side-band frequencies.

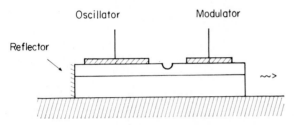

FIG. 3.24. Schematic section of the double diode structure used in frequency modulation experiments (Fenner, 1964)

3.8.3 *Frequency modulation*

Frequency modulation of a gallium arsenide laser can be achieved by modulating the refractive index of the cavity material. This can be achieved by coupling a laser to an ultrasonic transducer and in this way it has been possible to modulate a laser at 2·5 MHz obtaining a wavelength shift of 1·4 Å.

An alternative method of frequency modulation has been demonstrated by Fenner (1964) using a diode constructed as shown in Fig. 3.24. The first diode acts as the laser oscillator while electrons injected into the junction region of the second diode modulate the refractive index of this part of the cavity and hence frequency modulate the output of the laser.

Although both methods of frequency modulation have been demonstrated, they have not found any practical application since the modulated power is small and the techniques required complex. At the present time most applications involve some form of pulse code modulation or, in a few instances, microwave amplitude modulation.

3.9 Summary

One of the aims of this book is to show that gallium arsenide lasers have reached the stage of development where they are of practical as well as academic interest. It is therefore appropriate to conclude this chapter by summarizing some of the properties of typical lasers which might be of interest in practical applications (Table 3.7). The three examples chosen

TABLE 3.7. Basic properties of some typical devices operating at liquid nitrogen and room temperatures.

	Operating temperature			
	77	77	300	(°K)
Threshold current density	10^3	10^3	3×10^4	A/cm²
Typical devices	(a)	(b)	(c)	
Junction width	0·1	0·002	0·01	cm
Junction length	0·1	0·05	0·05	cm
Series resistance	0·01	0·1	0·2	ohm
Threshold current	10	1	15	A
Operating current	200	20	50	A
Maximum duty cycle	0·02	0·02	0·002	
Pulse length	5	5	0·1	μsec
Maximum repetition rate	4	4	20	kHz
Peak power	50	5	5	W
Mean power	1	0·1	0·01	W

have been confined to pulsed operation at liquid nitrogen and room temperatures since it is considered that these conditions are the most likely to be encountered in practice; devices such as these are readily obtainable. With this in mind the value of the various properties quoted are not the best values which have been reported in the literature but values which have been obtained with fair reproducibility.

The study of the gallium arsenide lasers has now reached the stage where the main features of their properties are quite well understood. However, the investigation of some of the details is still hampered by the problems of making uniform devices with accurately controlled impurity distributions.

In many applications of gallium arsenide lasers there is a considerable premium in operating the device at room temperature. However, it has been shown that the performance of the devices which are available is severely limited at room temperature, particularly as regards the dutycycle at which they will operate and the mean power which can be obtained from them. Consequently much of the present research and development in this field is aimed at reducing these limitations and until considerable progress has been made in this direction it is unlikely that the full potential of gallium arsenide lasers will be exploited.

References

Ahearn, W. E., and J. W. Crowe (1966). *I.E.E.E. Quantum Electronics Conference, Phoenix.*

Alyanovskii, V. N., V. S. Bagaev, Yu. N. Berozashvili and B. M. Vul (1966). *Fiz. Tverd. Tela.*, **8**, 1091. (*Soviet Phys. Solid State*, **8**, 871, 1966.)

Antonoff, M. M. (1964). *J. Appl. Phys.*, **35**, 3623.

Armstrong, J. A., M. I. Nathan and A. V. Smith (1963), *Appl. Phys. Letters*, **3**, 68.

Armstrong, J. A., and A. W. Smith (1964). *Appl. Phys. Letters*, **4**, 196.

Biard, J. R., W. N. Carr and B. S. Reed (1964). *Trans. AIME*, **230**, 286.

Broom, R. F., C. H. Gooch, C. Hilsum and D. J. Oliver (1963). *Nature*, **198**, 4878.

Broom, R. F. (1968). *J. Quantum Electronics*, **4**, 135.

Burns, G., F. H. Dill and M. I. Nathan (1963). *Proc. I.E.E.E.*, **51**, 947.

Carlson, R. O. (1966). *J. Appl. Phys.*, **38**, 661.

Clifton, M., and P. R. Debye (1965). *Appl. Phys. Letters*, **6**, 120.

Cooper, D. P., C. H. Gooch and R. J. Sherwell (1966). *J. Quantum Electron.*, **2**, 329.

Crowe, J. W., and R. M. Craige (1964). *Appl. Phys. Letters*, **5**, 72.

Dobson, C. (1966). *Brit. J. Appl. Phys.*, **13**, 187.

Dousmanis, G. C., C. W. Mueller and H. Nelson (1963). *Appl. Phys. Letters*, **3**, 133.

Dousmanis, G. C., H. Nelson and D. L. Staebler (1964). *Appl. Phys. Letters*, **5**, 174.

Dousmanis, G. C., and E. F. Gross (1966). *Proc. I.E.E.E.*, **54**, 998.

Dyment, J. C. (1967). *Appl. Phys. Letters*, **10**, 84.

Dyment, J. C., and L. A. D'Asaro (1967). *Appl. Phys. Letters*, **11**, 207.

Dyment, J. C., and J. E. Ripper (1968). *J. Quantum Electron.*, **4**, 155.

Dyment, J. C., and T. H. Zachos (1968). *J. Appl. Phys.*, **39**, 2923.

Engeler, W. E., and M. Garfinkel (1964). *J. Appl. Phys.*, **35**, 1734.

Feinleib, J., S. Groves, W. Paul and R. Zallen (1963). *Phys. Rev.*, **131**, 2070.

Fenner, G. (1963). *J. Appl. Phys.*, **34**, 2955.

Fenner, G. (1964). *Appl. Phys. Letters*, **5**, 198.

Fenner, G. (1967). *Solid-State Electron.*, **10**, 753.

Garfinkel, M., and W. E. Engeler (1963). *Appl. Phys. Letters*, **3**, 178.

Goldstein, B. S., and R. M. Weigand (1965). *Proc. I.E.E.E.*, **53**, 195.

Gonda, T. (1965). *J. Quantum Electron.*, **1**, 159.

Gooch, C. H. (1965). *Phys. Letters*, **16**, 5.

Gooch, C. H. (1966). *Proc. GaAs Symp., Reading*, p. 62.

Kingsley, J., and G. Fenner (1963). *Bull. Am. Phys. Soc.*, **8**, 87.

Konnerth, K. (1965a). *Proc. I.E.E.E.*, **53**, 398.

Konnerth, K. (1965b). *I.E.E.E. Trans. Electron. Devices*, **12**, 506.

Konnerth, K., and C. Lanza (1964). *Appl. Phys. Letters*, **4**, 120.

Kressel, H. (1968). *J. Quantum Electron.*, **4**, 176.

Kressel, H., and N. E. Byer (1969). *Proc. I.E.E.E.*, **57**, 25.

Ludman, J. E., and K. M. Hergenrother (1966). *Solid-State Electron.*, **9**, 863.

Malstrom, L. D., J. J. Schlikmann and R. H. Kingston (1964). *J. Appl. Phys.*, **35**, 248.

Marinace, J. C. (1964). *IBM J. Res. Develop.*, **8**, 543.

Marple, D. J. F. (1964). *J. Appl. Phys.*, **35**, 1241.

Miller, R. C., F. M. Ryan and R. R. Emtage (1964). *Proc. Symp. Rad. Recomb., Paris*, p. 209.

Nathan, M. I., A. B. Fowler and G. Burns. (1963). *Phys. Rev. Letters*, **11**, 152.

Nelson, D. F., M. Gershenzon, A. Askin, L. A. D'Asaro and J. C. Sarace (1963). *Appl. Phys. Letters*, **2**, 182.

Nelson, H. (1967). *Proc. I.E.E.E.*, **55**, 1415.

Nelson, H., J. Pankove, F. Hawrylo and G. C. Dousmanis (1964). *Proc. I.E.E.E.*, **52**, 1360.

Pilkuhn, M., and H. Rupprecht (1963). *Proc. I.E.E.E.*, **51**, 1243.

Pilkuhn, M. H., and G. T. Guetler (1968). *J. Quantum Electron.*, **4**, 132.

Pilkuhn, M. H., H. Rupprecht and J. Woodall (1965). *J. Quantum Electron.*, **1**, 184.

Ryan, F. M., and R. C. Miller (1963). *Appl. Phys. Letters*, **3**, 162.

Seany, R. J. (1967). Unpublished results.

Sorokin, P. P., J. D. Axe and J. R. Lankard (1963). *J. Appl. Phys.*, **34**, 2553.

Stern, F. (1966). *Phys. Rev.*, **148**, 186.

Stevenson, M. J., J. D. Axe and J. R. Lankard (1963). *IBM J. Res. Develop.*, **7**, 155.

Sturge, M. D. (1962). *Phys. Rev.*, **127**, 768.

Susaki, W. (1967). *J. Quantum Electron.*, **3**, 332.

Takamiya, S., F. Kitosawa and J. I. Nishizawa (1968). *Proc. I.E.E.E.*, **56**, 135.

Torrey, H. C., and C. A. Whitmer (1948). *Crystal Detectors, M.I.T. Radiation Laboratory Series.*

Vilner, L. D. (1962). *Opt. i Spectroskopiya*, **12**, 437. (*Opt. Spectra U.S.S.R.*, **12**, 240, 1962).

Winogradoff, N. N., and H. K. Kessler (1964). *Solid-State Commun.*, **2**, 119.

Yonezo, H., A. Kawaji and Y. Yasovka (1968). *Solid-State Electron.*, **11**, 129.

4

The Preparation and Properties of Gallium Arsenide

M. C. ROWLAND

Services Electronics Research Laboratory, Baldock, Hertfordshire

Contents

4.1   Historical

The III–V semiconductors are a group of compounds formed by combining elements from group IIIb and group Vb of the periodic table. These form definite compounds having fixed 1 : 1 ratio between the elements. Goldschmidt (1929) showed that these III–V compounds were structural analogues of the group IV elements crystallizing in a form similar to the diamond lattice. Interest increased rapidly after Welker (1952) showed that not only were these compounds structurally similar to the group IV elemental semiconductors but were also semiconductors themselves. A more detailed account (Welker, 1954) of the properties of this family of semiconductors gave information on the energy gap, electron mobility and optical absorption of many of these compounds including gallium arsenide.

The first radiative recombination in GaAs was observed by Braunstein (1955) who obtained emission from minority carriers injected into bulk material from point contacts and also from silver paste blocking contacts. Further work on the crystal growth of GaAs produced large, pure samples which enabled the study of the physical properties and technology of the material to proceed rapidly. Reports of high internal efficiencies for radiative recombination at p–n junctions by Keyes and Quist (1962) and by Pankove and Berkeyheiser (1962) led finally to the first injection lasers by Hall and coworkers (1962), by Nathan and coworkers (1962) and Quist and coworkers (1962).

Since then a large amount of theoretical and practical effort has been devoted to the study of GaAs semiconductor lasers and this has resulted in improved power outputs and higher operating temperatures. These improvements are directly related to improvements in crystal growth and, in particular, to the development of low temperature liquid phase systems.

4.2 Crystal structure and properties

4.2.1 Crystal structure

Like most of the III–V compounds GaAs crystallizes in the zinc-blende lattice in which one atom is at the centre of a regular tetrahedron whose four corners are atoms of the other kind. This structure is very similar to

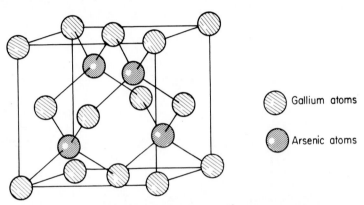

Gallium atoms

Arsenic atoms

FIG. 4.1. Cubic structure of the gallium arsenide zinc-blende lattice. The tetrahedral bonding of the atoms is indicated

the diamond lattice. However, in the diamond lattice all the sites are occupied by the same atoms, whereas in the zinc-blende lattice the atoms are alternately A and B. The typical cubic cell of GaAs is shown in Fig. 4.1, a face-centred cubic lattice of gallium atoms with arsenic atoms positioned on the body diagonals. These arsenic atoms also lie on a face-centred cubic lattice displaced relative to the gallium lattice by one-quarter the body diagonal of the cube.

4.2.2 Defects

In the GaAs crystal lattice there exists the possibility of a large number of single and multiple defects. Single defects are gallium or arsenic vacancies or interstitials. Multiple defects such as gallium–arsenic or gallium–gallium divacancies and vacancy–impurity complexes may be formed depending on the method of preparation and sample treatment.

Fuller, Wolfstirn and Allison (1967) studied defect centres in GaAs and suggested the presence of 10^{17} cm^{-3} defects in normal boat-grown material. After heat treatment at 450 to 800°C this could increase to 10^{18}–10^{19} cm^{-3} (Blanc, Bube and Weisberg, 1964).

Very few studies have been made on the effect of crystal defects, such as dislocations or stacking faults, on the performance of GaAs devices. Laser junctions were investigated by Abrahams and Buiocchi (1966) who showed a direct correlation between decorated interface dislocations and the sites of first laser emission. McCarthy (1966) was able to associate microplasma breakdown with dislocations in reverse biased epitaxial *p–n* junctions. Williams (1967) showed that substrate dislocation density effected the electroluminescent efficiency of epitaxial GaAs diodes.

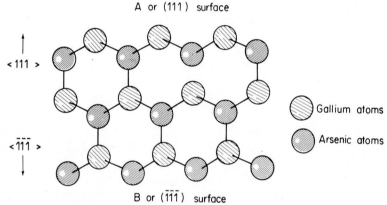

FIG. 4.2. The structure of the gallium arsenide lattice on a (110) plane showing the distinctive (111) and ($\bar{1}\bar{1}\bar{1}$) surfaces

4.2.3 Symmetry

Examination of the zinc-blende structure shown in Fig. 4.2 shows that as the crystal is formed of alternate layers of A and B atoms in the [111] directions it lacks the inversion symmetry present in the diamond lattice. Due to this lack of symmetry the [111] directions form polar axes which result in distinctive (111) and ($\bar{1}\bar{1}\bar{1}$) faces as indicated. White and Roth (1959) distinguished and positively identified these two faces by special x-ray techniques and by etching. They showed that pyramidal etch pits produced at dislocations were developed only on the (111) face which is composed of gallium atoms, the ($\bar{1}\bar{1}\bar{1}$) face which is composed of arsenic atoms producing a smooth etched surface. This characteristic of the (111) surfaces in III–V compounds was also reported by Gatos and Lavine (1960).

4.2.4   Binding

Since the diamond and zinc-blende structures are so similar the electronic bonding might also be expected to be similar. It is well known that all materials which crystallize in the diamond-type lattice have purely covalent bonds. However, due to the difference between the atoms, the binding in the zinc-blende lattice, although predominately covalent, has a small ionic bond in addition. Cleavage studies (Wolff, 1962) have suggested that the degree of ionic nature in the GaAs bond is greater than 25%. The form and nature of the chemical bond in III–V semiconductors has been reviewed in detail by Folberth (1962).

4.2.5   Cleavage

Diamond cleaves most easily along (111) planes where the number of bonds per unit area is at a minimum and although GaAs will, with difficulty, cleave on the (111) planes the normal easy cleavage is on the (110) planes. Reference to Fig. 4.2 shows that the (111) planes are composed alternately of all gallium and all arsenic atoms and as these planes will have opposite charge, an electrostatic force will tend to prevent cleavage. On the other hand, (110) planes are all composed of equal numbers of Ga and As atoms in a chequer-board pattern so there will be no overall electrostatic force between the planes; in fact a shift in position of atoms in the (110) plane will allow them to assume a repel position making them easier to separate.

4.3   Crystal growth from the melt

Many of the characteristics and properties of GaAs lasers such as threshold current and efficiency are dependent on the nature of the material used for the fabrication of the devices. This, therefore, places considerable demands on the technology of crystal growth; in particular, the control of chemical purity and crystalline perfection. Ideal laser material would be large, single crystal ingots uniformly doped with a chosen impurity. The material should have no other chemical impurities or crystalline defects as these can cause non-radiative recombination or radiation at an undesired wavelength, both decreasing the efficiency of the laser. Although it is possible to grow large single crystals, chemically pure GaAs and GaAs free from crystalline defects, each requirement is, at present, fulfilled by a different growth technique, so that in general the system used will be a compromise based on the material characteristics required.

Most of the standard techniques for preparing metals and the elemental semiconductors have been adapted for the growth of GaAs; these include conventional growth from a melt, crystal pulling, zone refining and epitaxial growth. However, because of the high dissociation vapour pressure of arsenic at the melting point these systems are in general more complicated when used for GaAs.

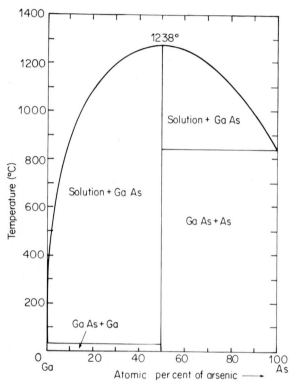

Fɪɢ. 4.3. The temperature–composition phase diagram for the system gallium–arsenic (after Koster and Thoma, 1955)

4.3.1   Phase equilibrium

The phase diagram for the system Ga–As as determined by Koster and Thoma (1955) shows, as in Fig. 4.3, that the stoichiometric compound is formed by equal atomic fractions of the two elements and has a melting point of 1238°C. Straumanis and Kim (1965) have determined the phase extent of GaAs using measurements of the lattice constant and density of bulk samples. They concluded that the phase extends at least from 49·998 to 50·009 atomic per cent arsenic. The equilibrium vapour pressure of the system along the GaAs liquidus has been given by Thurmond (1965) and by Arthur (1967). They have determined the vapour pressures of the different vapour species and these are shown in Fig. 4.4. At the melting point the major volatile species are As_4 and As_2 with partial pressures of 0·648 and 0·328 atm, respectively. The corresponding vapour pressures of

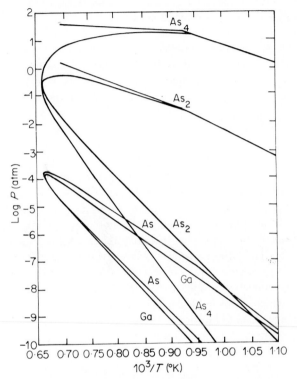

FIG. 4.4. Equilibrium vapour pressures of As, As_2, As_4 and Ga along the liquidus as a function of T^{-1}. Pressures of As_2 and As_4 over pure solid and liquid arsenic are also shown (after Arthur, 1967)

the monatomic species arsenic and gallium are less than 10^{-4} atm. A melt of GaAs produces a dissociation vapour pressure of 0·976 atm at the melting point and in order to produce a stoichiometric solid all melt-growth systems must have a means of maintaining and controlling this vapour pressure of arsenic over the melt during solidification. Brice (1967) suggests that this excess pressure controls the number of arsenic vacancies in melt-grown GaAs and that this affects the efficiency of electroluminescent devices made from the material.

Again referring to the phase diagram and the vapour-pressure diagram it can be seen that growth at temperatures well below the melting point can be achieved in gallium-rich or arsenic-rich liquid phases, and these systems will result in lower or higher arsenic vapour pressures, respectively. Solution-grown techniques have produced new methods for the production of laser material and they will be described in the section on epitaxy.

4.3.2   Arsenic vapour pressure control

Despite the high vapour pressure of arsenic at the melting point all the standard semiconductor techniques of zone melting, float-zoning, melt growth and crystal pulling have been adapted for the preparation of stoichiometric GaAs. However, because of the high arsenic vapour pressure at the melting point these systems all have to be modified to produce some means of controlling the vapour pressure of arsenic above the melt. This generally requires the use of a sealed quartz system with its coldest point regulating the arsenic vapour pressure during the growth. Two methods are available for this. The first employs excess arsenic condensed at the minimum temperature point of the system to supply the required vapour pressure. The second method uses only sufficient arsenic to maintain 1 atm vapour pressure in the system with the complete system above the condensation point. The latter method has the advantage that the controlled arsenic vapour pressure is much less sensitive to temperature variations, because the pressure varies linearly with temperature whereas in the first method the vapour pressure of arsenic over solid arsenic varies exponentially with temperature (Thurmond, 1965).

The fact that the dissociation vapour pressure of arsenic at the melting point of GaAs is 0·976 atm is indeed fortunate as it enables GaAs to be synthesized in normal walled quartz apparatus. The internal arsenic vapour pressure counteracts the external atmospheric pressure which would otherwise cause the quartz ampoule to collapse, as it is already softening, at the temperatures used.

4.3.3   Purity of the elements

The first workers on GaAs made considerable efforts to purify the elements before reacting them to produce the compound. Methods for the purification of both gallium and arsenic are now so well developed in industry that further purification in the laboratory is unnecessary and is more likely to contaminate than improve the product. A review of the steps necessary for the purification of gallium is given by De La Breteque (1962) and for arsenic by Blum (1962). With both elements available at greater than 99·999% purity the analysis, identification and removal of the residual impurities become extremely difficult. Mass spectrographic analysis of gallium shows only traces of magnesium, aluminium, silicon, calcium, manganese and iron all at the limit of detectability, less than 0·05 p.p.m. Detectable impurities in arsenic are magnesium, silicon and copper at less than 0·1 p.p.m. It is obvious, therefore, that all handling, weighing and storage of the elements need to be carried out in clean, dust-free conditions avoiding contact with impure reagents, metal containers or tweezers. Arsenic should be stored under vacuum as the element oxidizes readily in air.

4.3.4 Melt-growth systems

The growth of GaAs from a stoichiometric melt held in a crucible is the oldest and simplest method of crystal growth. The techniques for melt growth may be differentiated by the method used to solidify the melt. In the systems which use a completely liquid charge we have the Bridgman (1925), Stober (1925) and Czochralski (1917) techniques. The first two systems are practically identical and differ only in the method by which the temperature gradient is passed through the melt. Bridgman systems move the crucible relative to the furnace whereas the Stober or static-freeze technique employs a stationary system and programmes a temperature gradient through the melt. The third system which employs a completely liquid melt is the Czochralski pulling technique.

Systems employing a small molten zone in an otherwise solid ingot are the horizontal zone melting system (Richards, 1960) and the zone refining techniques described by Cunnell, Edmond and Harding (1960).

[i] *Horizontal Bridgman method.* A typical system for the growth of GaAs described by Cunnell, Edmond and Harding (1960) is shown in Fig. 4.5. The quartz ampoule is about 80 cm long and 3 cm diameter. The melt is contained in a quartz boat which is normally sand-blasted or carbon-coated on the inner surface to prevent wetting and adherence, which would otherwise occur. If wetting does occur the solid ingot shatters on cooling owing to the different thermal expansions of GaAs and quartz. The ampoule is assembled with the boat containing the weighed elements with a small excess of arsenic and the required dopant. Provided the dopant is not volatile, the ampoule is joined onto a vacuum system and baked at 250°C for several hours to remove any arsenic trioxide. Volatile dopants such as zinc or sulphur must be added after this step.

The evacuated ampoule is then sealed off and placed in the ceramic tube, over which the three-zone furnace can be moved on rails. The low-temperature furnace zone maintains the atmosphere of arsenic vapour throughout the growth and is normally controlled by a thermocouple strapped onto the cold end of the quartz ampoule. Two platinum-wound furnaces provide the hot zone at 1250°C and a temperature gradient through which the melt is moved. The gradient shape and steepness is extremely important as it controls the solid–liquid interface geometry which is the main factor in determining the crystallinity and perfection of the solid phase. Typical gradients used in these systems are 5–10°C/cm and growth rates vary from 0·5 to 3·0 cm/h. GaAs ingots of 100 g weight with a cross section of 1 cm² have been grown; these usually contain large single crystals and are occasionally completely single.

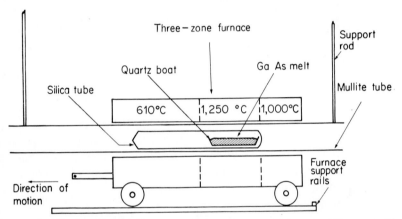

Fig. 4.5. A horizontal Bridgman system for the preparation of gallium arsenide

The main disadvantages of this method arise from the vibration associated with the mechanical movements and the difficulty of accurately controlling the temperature of the arsenic cold point throughout the growth cycle. Low arsenic vapour pressures cause a gallium-rich melt which results in polycrystalline ingots with free gallium in veins between the grains. High arsenic vapour pressures lead to an arsenic-rich melt. Solidification of the stoichiometric solid then causes an accumulation of excess arsenic adjacent to the interface. This liquid therefore has a higher dissociation vapour pressure than the ambient and excess arsenic boils off. If growth rates are slow then this may not be noticed but at fast growth rates the resulting crystals can be quite sponge-like with many bubbles and voids trapped in the crystals. As might be expected, dislocation densities in the vicinity of these macroscopic defects and gallium inclusions are much higher than in the bulk crystal. Good crystals can have dislocation densities between 10^3 and 10^4 cm^{-2}.

[ii] *Static freeze method.* Several of the disadvantages of the horizontal Bridgman technique can be overcome by using a stationary system and a moving temperature gradient. Vibration is no longer a problem and easier control of the temperatures, particularly that of the arsenic reservoir, results with a fixed furnace.

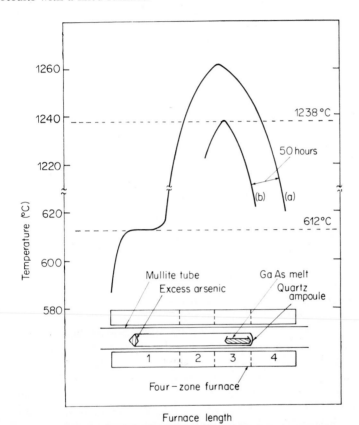

Fig. 4.6. Static freeze system for the preparation of gallium arsenide. Profiles (a) and (b) represent the temperature conditions at the beginning and end of the 50-hour growth cycle

One difference in the quartz ware used for these two systems is the design of the boat. In the static freeze system the length of the boat is determined by the distance over which the gradient, having a maximum temperature of 1260°C, can be moved before the maximum falls to 1240°C. Fig. 4.6 shows schematically the system in use at this laboratory. Each high-temperature platinum furnace is controlled by a thyristor controller to an accuracy of $\pm 1 \cdot 0°$. The excess arsenic condensed at the cold end of the ampoule is controlled at a temperature of $612 \pm 0 \cdot 5°C$ to give $1 \cdot 0$ atm arsenic vapour pressure. This is slightly above the $0 \cdot 976$ atm dissociation vapour pressure required for a stoichiometric melt but enables the quartz ampoule to withstand $1260°C$ throughout the growth cycle without collapsing, and provided growth rates are slow this excess pressure does not affect the crystalline perfection of the ingots.

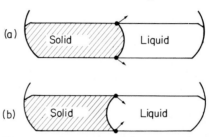

FIG. 4.7. The effect of the interface shape
on the growth of single crystals
(a) Convex solid face immediately grows
out stray crystal nuclei
(b) Concave solid face allows stray
crystals to propagate throughout the
remaining crystal length

The gradient is determined by the ratio of power fed to the two furnace zones 3 and 4. As Brice (1965) points out the ideal temperature profile for a melt-growth system would have a low maximum temperature to prevent container reactions, a high gradient in the liquid to prevent constitutional supercooling and a low gradient in the solid to eliminate strain.

In all melt-growth systems, the shape of the solid–liquid interface controls the crystallinity of the ingot. Hurle (1962) describes the effect of convex and concave interfaces on the crystal structure of the resulting ingot. Reference to Fig. 4.7 shows that, provided growth occurs normal to the interface, the convex solid face ensures that spurious crystal nucleations tend to grow out and are eliminated, whereas with a concave face they propagate into the bulk of the crystal and produce polycrystalline ingots.

The system is operated with a maximum temperature of $1260°C$ and a temperature gradient of $4°/cm$ at the interface. Rates of freezing between $0 \cdot 2$ and $0 \cdot 6$ cm/hour have been used to produce largely single crystal ingots. The slower rate prevents a large volume of the melt supercooling and spontaneously freezing, and therefore produces more single crystal ingots. About 60% of all ingots are largely single crystals. Fig. 4.8 shows examples of doped and undoped single crystals, each weighing 150 g, produced by the static freeze technique.

Electrical assessment of the carrier concentration down the length of an ingot (Skalski, 1962) can show large variations. The commonly employed dopants all have very small distribution coefficients (Willardson and Allred, 1967) which result in a gradually increasing dopant concentration in the melt and consequently in the solid as freezing proceeds, the last frozen end being more heavily doped.

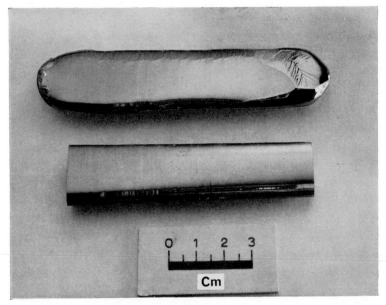

Fig. 4.8. Single crystal GaAs ingots grown by the static freeze technique

Typical electrical measurements made on static freeze ingots are shown in Table 4.1. This table shows the weight of various dopants added to the melt and the resulting Hall coefficient R_H, resistivity ρ and mobility μ of samples taken from each end of the ingot. The dopants used for lasers, in this laboratory, are chiefly silicon and tellurium. Chromium-doped semi-insulating and zinc-doped *p*-type crystals are required for certain epitaxial systems and details of these are included in the table.

[iii] *Horizontal zone melting.* This technique was originally developed for the purification of germanium (Pfann 1952) and described by Pfann and Olsen (1953). The method was adapted for the production of single crystal GaAs by Richards (1960).

TABLE 4.1. Typical electrical assessments of GaAs crystals grown by the static freeze technique.

| Dopant | | Electrical measurements at room temperature | | | | | |
| | | Ingot head | | | Ingot tail | | |
Impurity element	Weight mg/100 g GaAs	R_H (cm³/coulomb)	ρ (Ω cm)	μ (cm²/V sec)	R_H (cm³/coulomb)	ρ (Ω cm)	μ (cm²/V sec)
Si	20	3·0	$1·2 \times 10^{-3}$	$2·5 \times 10^{3}$	1·4	$1·0 \times 10^{-3}$	$1·4 \times 10^{3}$
Te	20	14·5	4×10^{-3}	$3·6 \times 10^{3}$	3·6	$1·7 \times 10^{-3}$	$2·1 \times 10^{3}$
	80	4·8	$1·8 \times 10^{-3}$	$2·7 \times 10^{3}$	1·0	$5·0 \times 10^{-4}$	$2·0 \times 10^{3}$
Zn	7·5	18·6	$1·0 \times 10^{-1}$	$1·9 \times 10^{2}$	10·6	$6·4 \times 10^{-2}$	$1·7 \times 10^{2}$
	20·0	7·34	$5·7 \times 10^{-2}$	$1·3 \times 10^{2}$	6·7	$3·8 \times 10^{-2}$	$1·8 \times 10^{2}$
Cr	70	4×10^{9}	5×10^{6}	8×10^{2}	$3·5 \times 10^{10}$	6×10^{7}	6×10^{2}

Fig. 4.9 shows a typical system in which a polycrystalline ingot is held at 900°C in a quartz boat. The movable insulated section of the furnace raises the temperature of a small zone of the ingot above its melting point and this is transversed the length of the ingot at 1·5 cm/h. The liquid zone is maintained stoichiometric by the arsenic reservoir held at 610°C.

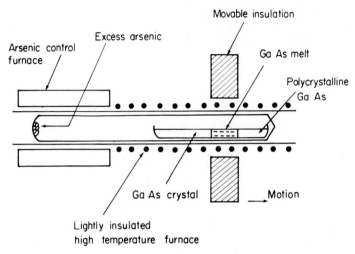

FIG. 4.9. Horizontal zone melting system

Crystal growth on chosen orientations can be obtained by initiating the growth from a seed crystal placed at the end first melted. Experimentally the ⟨111⟩ directions are found to give the most perfect crystals. Uniformly doped crystals can be produced by including impurities having small distribution coefficients.

Dislocation densities as low as 10 counts/cm² have been obtained by extremely careful control of the temperature gradients; more usually, dislocation densities are between 10^2 and 10^4 cm^{-2}.

[iv] *Float-zone techniques.* In order to avoid the difficulty of finding a suitable non-reactive crucible for the growth of GaAs, floating-zone techniques developed for silicon were adapted to GaAs by Whelan and Wheatley (1958). In the simplest system the presynthesized rod of GaAs is held vertically in two chucks and rotated whilst a small molten zone is passed through the bar. Impurities tend to remain in the liquid phase and are moved to one end of the ingot, resulting in a purification during each pass. Cunnell and Wickham (1960) describe in detail a quartz system capable of producing single crystal rods of 0·6–1·0 cm diameter and 10 cm

long. After more than a few passes the GaAs becomes semi-insulating, probably due to compensation caused by oxygen contamination in the melt. Doping in this system is not practical and such crystals are not suitable for laser work. Dislocation densities for these zone-refined rods are typically greater than 10^4 cm^{-2}.

4.3.5   Crystal pulling from the melt

In order to use the Czochralski technique for the growth of GaAs many modifications have to be made to the simple system used for the elemental

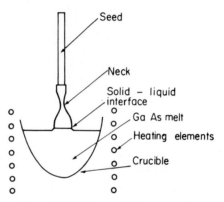

FIG. 4.10. Schematic diagram of a
Czochralski crystal puller

semiconductors germanium and silicon. Once again the high vapour pressure of arsenic at the melting point needs accurate control and since the system requires rotation and vertical movement it adds considerable difficulties to the puller design.

The fundamental parts of a Czochralski puller are shown in Fig. 4.10, the essential parts being the crucible, a means of heating and an orientated seed crystal.

In operation, the orientated, single crystal seed rotating at a fixed speed is allowed to melt back in the liquid to ensure good wetting. The melt temperature is then lowered slowly and as the GaAs seed crystal grows it is raised slowly to produce a narrow neck region. This narrow neck region is then withdrawn continuing the small diameter so that dislocations present in the seed, which are not parallel to the growth direction, grow out at the walls of the neck. The main crystal is then pulled, its shape being controlled by the programmed temperature of the crucible and the vertical withdrawal rate. Since the solid–liquid interface is not in contact with the crucible and the radial and vertical temperature distributions can be controlled accurately, this technique can grow dislocation-free crystals making it unique among melt-growth systems.

[i] *Sealed systems.* To operate a Czochralski puller for GaAs the whole system must be maintained above the arsenic reservoir temperature of 610°C. Gremmelmaier (1956) produced a completely sealed quartz system, transmitting rotation and vertical motion into the system by external magnets coupled to a high Curie point alloy sealed into a quartz cylinder which fitted closely into the outer envelope to provide a bearing at the operating temperature.

Steinemann and Zimmerli (1963) have grown GaAs ingots using a modified Gremmelmaier system which operated without a solid arsenic reservoir. More recently Steinemann and Zimmerli (1967) have used this system to grow large, dislocation-free crystals of both doped and undoped GaAs. The experimental conditions for this were strict control of the stoichiometry, low-temperature gradients in the solid crystal and a flat horizontal liquid–solid interface. By growing a long thin neck 1–2 mm diameter some 10–20 mm long they were able to prevent any dislocations from the seed propagating to the grown crystal. Ingots (100 g) of undoped and selenium-, tellurium- or zinc-doped material (to 5×10^{18} carriers/cm³) were grown in the $\langle 111 \rangle$ directions easily and in the $\langle 110 \rangle$ and $\langle 100 \rangle$ directions with some difficulty.

[ii] *Liquid seal puller.* Richards (1957) produced a quartz apparatus which consisted of two parts joined by a liquid seal. The upper bell-shaped half dipping into a gallium pool held in a deep well in the lower half. Rotation and lifting were then possible. The whole apparatus and seal operated at 650°C and apart from some reaction between the gallium and arsenic, the seal remained unaffected during use.

[iii] *Syringe puller.* Moody and Kolm (1958) produced a simple system in which the arsenic loss was prevented by a precision-ground quartz piston in the outer quartz tube. A clearance of 12 μm formed a vacuum-tight syringe. An improvement by Cronin, Jones and Wilson (1963) used a boron nitride rod in a boron nitride bearing and added a pressure of argon behind the seal to prevent arsenic diffusion up the joint.

[iv] *Liquid encapsulation.* An interesting technique for preventing the loss of a volatile component described by Metz, Miller and Mazelsky (1962) for PbSe and PbTe employed a liquid cover over the surface of the melt to prevent vapour loss. Mullin, Straughan and Brickell (1965) called this technique 'liquid encapsulation' and showed that the technique was useful for the III–V compounds and could prevent the loss of the group V

element. Bass and Oliver (1967) described the properties of doped and undoped GaAs produced by this technique using boric oxide as an encapsulant. Fig. 4.11 shows the system used. A water-cooled container filled with an inert gas was provided with standard seals for rotation and

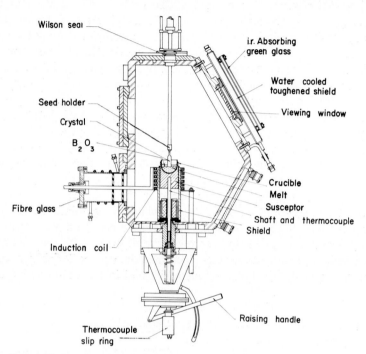

FIG. 4.11. Part sectional view of crystal pulling furnace (after Bass and Oliver, 1967)

lifting. The induction heated graphite susceptor held a quartz crucible from which the crystals were pulled. Provided that the B_2O_3 was thoroughly dried the attack on the crucible was small and the B_2O_3 could be used for several successive pulls. A slight disadvantage of this technique is that the GaAs cannot be synthesized in the apparatus but must first be produced in a Bridgman or static freeze system.

GaAs single crystals of 90 g were pulled on the $\langle 111 \rangle$, $\langle \bar{1}\bar{1}\bar{1} \rangle$, $\langle 100 \rangle$ and $\langle 110 \rangle$ directions. Using $\langle 111 \rangle$ orientated seeds, dislocations could be removed completely from the initial part of the ingot by growing a long thin neck. Single crystals doped with tellurium or zinc with carrier concentrations up to 2×10^{19} cm^{-3} were produced. Fig. 4.12 shows a typical ingot weighing 80 g pulled on $\langle 111 \rangle$ seed crystal.

Contamination from boron was very low and was determined as 0·01 p.p.m. by weight from a mass spectrographic analysis. Silicon was the major impurity at 1 p.p.m. by weight. Other impurities tended to oxidize and dissolve in the liquid boric oxide.

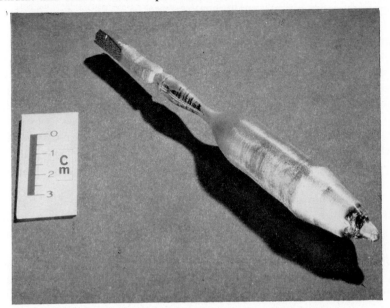

FIG. 4.12. A pulled crystal (after Bass and Oliver, 1967)

4.3.6 Contamination during melt growth

Most GaAs is produced and processed in quartz boats in a quartz apparatus as this material provides a suitable sealed system which will stand the required temperatures and pressures. There are a number of possible sources of contamination in these systems, the chief being the attack of the melt on the quartz boat. Silicon contamination from the boat above 5×10^{16} atoms/cm^3 has been found by Ekstrom and Weisberg (1962). This results from a series of chemical reactions between gallium and the quartz and these have been studied in detail by Cochran and Foster (1962) who have identified the following equilibrium reactions as important in boat-grown synthesis.

$$4 \, Ga_{(1 \, or \, GaAs)} + SiO_{2(s)} \rightleftharpoons Si_{(in \, Ga)} + 2 \, Ga_2O_{(v)} \tag{4.1}$$

$$Si_{(in \, Ga \, or \, GaAs)} + SiO_{2(s)} \rightleftharpoons 2 \, SiO_{(v)} \tag{4.2}$$

$$2 \, Ga_{(1 \, or \, GaAs)} + SiO_{2(s)} \rightleftharpoons Ga_2O_{(v)} + SiO_{(v)} \tag{4.3}$$

$$3 \, Ga_2O_{(v)} + As_{4(v)} \rightleftharpoons Ga_2O_{3(s)} + 4 \, GaAs_{(s)} \tag{4.4}$$

Equation (4.1) shows the mechanism by which the GaAs can be contaminated with silicon. This reaction continues until the silicon activity in the liquid is so high that further silicon produced reacts with the quartz to produce silicon monoxide as shown in Eqn. (4.2). The combination of these gives the overall equilibrium reaction (4.3). Once equilibrium is established there is a fixed level of silicon and this can only increase if the volatile oxides are removed. This can occur if the gallium suboxide dissolves in the melt or reacts at the cold end of the system as in Eqn. (4.4), resulting in an overall transport of GaAs. Cochran and Foster suggested that the silicon contamination could be suppressed by deliberately introducing a partial pressure of Ga_2O into the system, from Ga_2O_3 or an oxygen overpressure, thereby suppressing (4.1) and (4.3). They also suggested removing the cold surfaces as far as possible from the hot reaction zone, reducing the tube diameter between the melt and the arsenic reservoir and minimizing the volume of the free space in the reaction zone to reduce the silicon contamination.

Ainslie, Blum and Woods (1962) showed that the addition of oxygen in the 10–20 torr range reduced the residual carrier concentration in the material and gave increased electron mobilities. Woodall and Woods (1966) have studied the resistivities and residual carrier concentrations of GaAs made with oxygen overpressures, producing material with resistivities in the range 0·5–1000 Ω cm by subsequent heat treatment. Woodall (1967) showed that the final silicon carrier concentration is fixed by the crystal growth rate, the system geometry and the initial Ga_2O pressure.

Knight (1961) produced GaAs in carbon-coated boats which had silicon impurities up to 10^{18} cm^{-3} liberated from the quartz by the reaction:

$$SiO_{2(s)} + 2 \, C_{(s)} \rightleftharpoons 2 \, Si_{(s)} + 2 \, CO_{(v)} \tag{4.5}$$

Hara and Akasaki (1966) have used vitreous carbon boats and compared the resulting ingots with similar ingots produced in quartz boats. Silicon levels of 6×10^{16} cm^{-3} were found in quartz and 5×10^{17} cm^{-3} in carbon boats. The absence of a melt–quartz interface prevents reaction (4.2) reaching equilibrium and therefore allows the silicon impurity level to rise uncontrollably.

Stearns and McNeely (1966) studied the effect of various boat materials on the residual silicon contamination, they found a carrier concentration of $2·4 \times 10^{16}$ cm^{-3} in GaAs synthesized in quartz but with the addition of 3·0 torr of oxygen an identical ingot measured $4·8 \times 10^{14}$ carriers/cm^3. An ingot produced in a boron nitride boat had a carrier concentration of 6×10^{17} cm^{-3}, the addition of 2·6 torr oxygen produced an ingot with $1·9 \times 10^{18}$ carriers/cm^3. With boron nitride the absence of any melt in contact with quartz prevented the formation of an equilibrium concentration of silicon in the liquid phase. The silicon concentration, therefore, increases continuously during processing. The addition of oxygen to the boron nitride growth increased the silicon contamination by the following reaction, which allows silicon to be transported from the quartz ampoule walls to the melt.

$$Ga_2O_{(v)} + 2 SiO_{2(s)} \rightleftharpoons 2 SiO_{(v)} + Ga_2O_{3(s)} \tag{4.6}$$

Sources of contamination, other than silicon, result from the impurities present in the quartz ware, impurities on the quartz surfaces derived from chemical reagents and water used in the cleaning steps. Atmospheric dust provides a varied source of contamination and can contain a great number of metallic elements. Even the small quantities of solid elements added to produce doped ingots if not themselves extremely pure can, in some cases, influence the crystal properties adversely.

4.4 Crystal growth from solution

4.4.1 Solution growth

Reference to the phase diagram of the Ga–As system, Fig. 4.3, shows that the liquidus temperature is reduced rapidly on either side of the stoichiometric composition. As already mentioned it is possible to grow stoichiometric GaAs from either excess arsenic or excess gallium phases provided that growth rates are slow, so that it becomes possible to produce solid crystals at temperatures well below the melting point. Growth from a gallium-rich phase has the additional advantage that the arsenic vapour pressure over the solution is very small, which simplifies the system considerably by removing the necessity for an arsenic reservoir. Once a melt is produced there are several alternative methods to obtain crystalline GaAs, the simplest being to cool slowly and allow free nucleation in the solution. This usually results in small platelets of GaAs which are not useful for lasers. This method of large volume solution growth has been described by Wolff, Keck and Broder (1954).

4.4.2 Travelling solvent method

It is possible to grow large crystals by arranging to feed the solution with solid GaAs at a higher temperature than the growing faces. This system has been termed the 'travelling solvent method' by Mlavsky and Weinstein (1963). The method has the advantage that as the melt is free of container restraint it is possible to grow out dislocations propagated from the seed crystal during growth. Wolff and Das (1966) grew GaAs by the travelling solvent method and showed that the grown material had lower dislocation densities than the seed crystal. This decrease was attributed to the formation of dislocation loops and by transformation to dislocations which propagated parallel to the plane of growth and terminated at the edges of the crystal. Weinstein, La Belle and Mlavsky (1966) showed that dislocations were eliminated in the first 0·15 mm of growth.

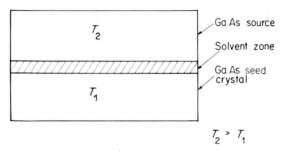

$T_2 > T_1$

FIG. 4.13. Travelling solvent method of crystal growth

Fig. 4.13 shows the basic system. The thin solution zone passes upwards under the force of the temperature gradient, solid material is removed from the face at T_2, transported through the liquid and deposited at the lower temperature face T_1. Gallium is usually chosen as the solvent although other elements such as tin or gold could be used. The usual growth direction is $\langle 111 \rangle$. The seed is single crystal whereas the feed material can be polycrystalline and doped either n- or p-type. Although large crystals can be produced growth rates are slow and the system is complicated when volatile dopants are used.

4.4.3   Liquid-phase epitaxy

The high concentration of impurity contamination in solid GaAs resulting from the reactions between the molten compound and the container can be greatly reduced, if not entirely removed, by growing the crystal at a low temperature. Two distinctly different systems have been developed to achieve this aim. The first employs a dilute solution of GaAs in gallium as the source of GaAs, the second uses a series of reversible chemical reactions between volatile gallium compounds and arsenic in the vapour phase to produce solid GaAs. When used to grow material for lasers, both systems employ a substrate to nucleate and grow suitable single crystal layers and are, therefore, termed epitaxial systems. The term 'epitaxial' is derived from the greek roots *epi* = upon, *taxis* = arrangement, and describes the growth of crystalline material from a fluid phase onto a seed crystal where the grown layer adopts exactly the crystal structure of the seed.

Nelson (1963) produced the first GaAs lasers from solution-grown material. His work showed that liquid-phase epitaxy had certain advantages for producing the highly doped extremely flat junctions needed for lasers.

The first reported (Broom, 1963) room-temperature operation of diffused GaAs lasers showed extremely high current thresholds necessitating very short pulses to avoid junction heating. Dousmanis, Nelson and Staebler (1964) showed how the room-temperature threshold could be lowered for lasers produced by liquid epitaxy. Winogradoff and Kessler (1964) compared diffused and epitaxial p–n junctions over the range from liquid nitrogen to room temperature. Their results showed that their epitaxial lasers had lower thresholds in the range -50 to $+20°C$. High efficiency lasers operating at room temperature have been reported by Nelson and coworkers (1964) who fabricated lasers from an n-type epitaxial layer on a zinc-doped substrate.

The liquid epitaxial techniques have several advantages for producing laser junctions. The systems are simple to make and operate, the growth cycle is rapid and the temperatures used are low. One added advantage of the technique is that the excess gallium present during growth prevents the formation of gallium vacancies (Williams and Blacknall, 1967) which is of particular interest in lasers which require material free of deep trapping levels.

A measure of the purity of the system can be obtained by growing undoped GaAs. The large gallium melt retains many of the impurities from the source material and the regrown material can have mobilities as high as 8500 cm²/V sec at 290°K and 106,000 cm²/V sec at 77°K (Andre and LeDuc, 1968) using boat-grown GaAs as source material. Studies by Panish and coworkers (1966) also showed a very efficient photoluminescence response in regrown layers which was twenty times higher than that obtained from melt-grown crystals. Goodwin, Gordon and Dobson (1968) produced very pure GaAs liquid epitaxial layers with an excess carrier concentration as low as 2×10^{14} cm^{-3} with a mobility of 7900 at room temperature and 52,000 cm²/V sec at 78°K. They used an AsCl$_3$ vapour-phase system to saturate a gallium source with GaAs and then transferred this to a horizontal liquid system for epitaxial growth.

[i] *Horizontal systems.* Pure undoped GaAs is dissolved at an elevated temperature in a suitable solvent such as tin, gallium, bismuth or lead and this saturated solution brought into contact with an orientated substrate.

The system is then cooled, resulting in the growth of an epitaxial layer on the substrate. Nelson (1963) used a horizontal furnace system flushed with pure hydrogen. In similar systems the saturated melt of GaAs in tin or gallium at one end of a carbon boat is moved onto the (100) orientated substrate at the other end by tipping the whole furnace. By increasing the temperature a few degrees at this point, material can be dissolved from the substrate providing an extremely flat, clean, accurately orientated (100) plane for the subsequent epitaxial growth which results when the system is cooled. The substrate is removed at 400°C and excess gallium removed with a P.T.F.E. blade, the small amount remaining being etched off in hot dilute HCl. Using a tin melt it is possible to produce layers using a maximum temperature of only 650°C which are doped as high as 2×10^{18} carriers/cm^3. As the (100) plane is a favourable growth plane, junctions can be produced in this plane and as these are perpendicular to the (110) cleavage planes lasers may be fabricated easily by sawing the slice into bars and then cleaving these bars on the (110) planes giving parallel-sided laser dice.

[ii] *Vertical systems.* More accurate control of the temperatures during the etch-back and regrowth parts of the process can be obtained by using the vertical systems described by Rupprecht (1967) and Williams and Blacknall (1967). As shown in Fig. 4.14, these systems use a quartz or boron nitride crucible and a vertical movable plate to hold the substrate. The substrate can easily be withdrawn from the melt at any time during the cooling cycle with a minimum amount of gallium attached. Horizontal systems do not give this facility as the melt tends to adhere strongly to the substrate surface and, in addition, the crust of GaAs crystals which forms over the melt prevents any movement. The substrate can be held in the cold portion of the furnace whilst the saturated solution is formed thereby preventing the loss of arsenic from the substrate which can occur in the horizontal system. As the crucible and substrate holder have thermocouples attached, the temperatures of both can be continuously monitored during the process. This is important, as Rupprecht (1967) showed that the flatness of the $p–n$ junctions is controlled by the temperature difference between the seed and saturated solution just prior to contact. The amount of etch-back is also controlled by this temperature difference. Flat junctions in his system resulted only when this difference was less than 7°C.

Growth rates are fast compared to other epitaxial growth techniques and can be as high as 20 μm per minute. Layer thickness is determined by the size of the melt, the temperature of operation and the temperature range of the cooling cycle. Solution-grown epitaxial layers are in general flat and featureless but poor temperature control during the etch-back and growth stages can cause irregular interfaces and poor surfaces. Strack (1967) suggested that the presence of a crucible wall close to the growth face improved the surfaces of the epitaxial layers.

[iii] *Liquid epitaxial lasers.* For laser junctions it is usual (Nelson and coworkers, 1964; Pilkuhn and Rupprecht, 1967) to grow *n*-type epitaxial layers on a *p*-type zinc-doped (100) substrate. These *n*-type layers are

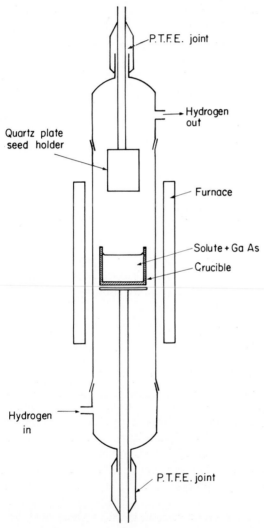

FIG. 4.14. A vertical liquid epitaxial system

usually tin or tellurium doped. Solutions of GaAs in tin operated at 650°C give carrier concentrations around 2×10^{18} carriers/cm³ whilst tellurium-doped gallium melts require a temperature of 900°C to reach this carrier concentration.

A melt of 15 g gallium, 2 g GaAs and 50 mg tellurium was used at this laboratory to grow a series of epitaxial layers on p-type boat-grown substrates. At 900°C the first run produced a layer with a carrier concentration of 4×10^{18} cm^{-3} which fell to 2×10^{18} cm^{-3} on the fourth run, each layer was about 100 μm thick.

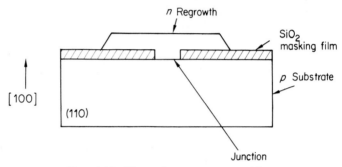

FIG. 4.15. Planar laser structure with a narrow junction grown by liquid epitaxy

Rupprecht (1967) used a vertical liquid epitaxial system to grow laser structures onto SiO$_2$ masked substrates. He obtained a narrow p–n junction with large contact areas by allowing the GaAs to overgrow the masked surfaces, thereby giving good electrical and thermal connections to a laser junction only 6 μm wide. Fig. 4.15 sketches the structure produced by this technique.

4.5   Crystal growth from the vapour phase

4.5.1   Vapour-phase epitaxy

Most vapour-phase systems, especially the open-tube methods, using either halogen or hydride transport, offer the crystal grower a great deal of control over the orientation, thickness and crystalline perfection of the epitaxial layers produced. They also offer an accurate control of doping levels and give the ability to produce chosen doping profiles to the extent of growing, in one process, a complete p–n laser junction. However, despite this versatility and the ease of operation, vapour-phase epitaxy has not achieved much prominence in the GaAs laser field. The main use, so far, has been in producing very pure GaAs for microwave devices. It has also been used extensively in gallium arsenide phosphide material growth where control over the arsenic to phosphorus ratio can be readily achieved to produce a homogeneous solid phase of predetermined composition.

[i] *Closed-tube growth.* The first vapour-phase preparation of GaAs was reported by Antell and Effer (1959) who prepared and transported GaAs down a temperature gradient in a closed quartz ampoule using iodine as the transporting agent. This system relied on the reversible reaction shown in (4.7)

$$6\ GaI_{(v)} + As_{4(v)} \rightleftharpoons 4\ GaAs_{(s)} + 2\ GaI_{3(v)} \tag{4.7}$$

Similar equations apply to the transport of GaAs using the other halogens bromine and chlorine. In each case all the reactants except GaAs are in the vapour phase. As the temperature is raised the reaction is driven from right to left to a new equilibrium position, transferring some solid GaAs into monohalide and free arsenic, these can then diffuse to the lower temperature seed zone. At this lower temperature the reaction reverses to deposit GaAs and reform the trihalide which can return to the source zone to react with more GaAs.

Pizzarello (1963) used this closed-tube system to produce epitaxial deposits of GaAs. With iodine as the transporting agent, good epitaxial layers were obtained on (111) GaAs seed crystals.

[ii] *Close-spaced systems.* Equation (4.8) shows a similar reversible reaction which has been used by Nicholl (1963) and Robinson (1963) to grow, in close-spaced systems, epitaxial GaAs on other III–V compounds and germanium. Gottlieb and Corboy (1963) studied the effects of temperature, flow rates and spacing in this system on the growth rate of GaAs on (111) GaAs seed crystals.

$$2\ GaAs_{(s)} + H_2O_{(v)} \rightleftharpoons Ga_2O_{(v)} + As_{2(v)} + H_{2(v)} \tag{4.8}$$

Fig. 4.16 shows schematically the system used. The graphite or molybdenum blocks support the seed and source slices and provide heat-sinks to maintain even temperatures. External infrared lamps provide a simple furnace system to give operating temperatures around 900°C and a source to seed difference of 30–40°C. The hydrogen atmosphere is kept moist by passing the incoming gas over ice at 0°C. The system achieves 95% deposition efficiencies because the closeness of the source and seed, about 0·5 mm, prevents the loss of the transporting species.

[iii] *Open-tube systems.* The open-tube growth systems for epitaxial growth, although more complex than the previous systems, are much more versatile and flexible in use and provide a better control of crystallinity and doping during the process. The first systems used HCl in a

hydrogen carrier gas stream transporting bulk GaAs from the high source temperature zone to a lower temperature seed zone (Williams and Ruehrwein, 1961). Goldsmith (1963) studied the effect of the seed orientation on the growth rates in this system. Effer (1965) used the HCl system to

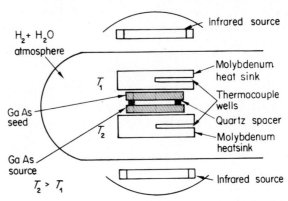

FIG. 4.16. Close-spaced system for growing epitaxial gallium arsenide

grow both selenium- and tin-doped GaAs epitaxial layers which could have a maximum concentration of 4×10^{18} carriers/cm^3. He showed that Eqns. (4.9) and (4.10) were the chief reactions occurring at the source and seed zones, respectively.

$$4\, GaAs_{(s)} + 4\, HCl_{(v)} \rightleftharpoons 4\, GaCl_{(v)} + As_{4(v)} + 2\, H_{2(v)} \tag{4.9}$$

$$6\, GaCl_{(v)} + As_{4(v)} \rightleftharpoons 4\, GaAs_{(s)} + 2\, GaCl_{3(v)} \tag{4.10}$$

The more recent vapour epitaxial systems are based on similar reactions using arsenic trichloride, gallium and hydrogen. These have the advantage that these reagents can be obtained extremely pure and, in addition, the epitaxial GaAs is synthesized in the system at low temperatures.

A typical system used is shown in Fig. 4.17. The two-zone furnace is controlled at around 800°C for the source zone and 750°C for the seed zone. The hydrogen stream of 50 to 100 cm^3/min can pass through or bypass the AsCl$_3$ container. The quartz ware is usually synthetic 'Spectrosil'.

$$4\, AsCl_{3(v)} + 6\, H_{2(v)} \rightarrow 12\, HCl_{(v)} + As_{4(v)} \tag{4.11}$$

Equation (4.11) shows the first reaction which occurs in the entrance to the furnace. The arsenic dissolves completely in the gallium source until it is saturated at 2·1 atomic % (Hall, 1963). During this period, which can be from 3 to 5 hours, no free arsenic deposits beyond the furnace but gallium is transported by the HCl as GaCl and GaCl$_3$, the dichloride is not stable at these temperatures, and these deposit at the exit end of the furnace.

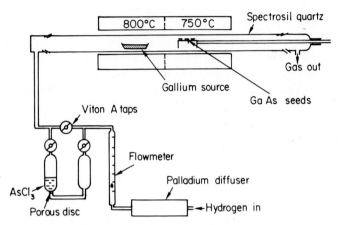

FIG. 4.17. Vapour phase epitaxial system

GaCl is unstable at room temperature and disproportionates to give free gallium and GaCl$_3$ according to Eqn. (4.12).

$$3\,GaCl_{(v)} \rightarrow GaCl_{3(s)} + 2\,Ga_{(s)} \qquad (4.12)$$

Once the source is saturated free arsenic immediately appears at the exit end of the furnace. The gallium source will now be covered with a thin crust of solid GaAs. If seeds are placed into the second zone epitaxial growth will occur. However, it is more usual to raise the temperature of the seed zone to about 900°C before inserting the seeds so that vapour etching takes place to produce a clean surface prior to deposition. After a few minutes the temperature is allowed to fall to 750°C when growth begins as shown in Eqn. (4.13).

$$6\,GaCl_{(v)} + As_{4(v)} \rightleftharpoons 4\,GaAs_{(s)} + 2\,GaCl_{3(v)} \qquad (4.13)$$

Growth rates depend on the source and seed temperatures, the AsCl$_3$ temperature, the hydrogen flow rates and seed orientation. Shaw and coworkers (1967) showed that with respect to substrate orientation the deposition rate was in the order

$$(111)A > (100) > (110) > (111)B.$$

Effer (1965) and Knight, Effer and Evans (1965) showed that very pure epitaxial layers could be produced using the $AsCl_3$, Ga, H_2 system. They were able to produce layers having an excess electron concentration of 2×10^{15} cm^{-3} with a mobility of 8,800 at room temperature and 40,000 cm^2/V sec at 77°K. More recently Whitaker and Bolger (1966) produced material with an excess carrier concentration of 1×10^{14} cm^{-3} and a mobility of 101,000 cm^2/V sec at 78°K.

Open-tube systems have been used to study the growth of GaAs through holes in SiO_2 masked-seed substrates. Tausch and LaPierre (1965) showed that although the GaAs overgrew the oxide film little nucleation took place on the film itself. Shaw (1966) etched holes in oxide-masked semi-insulating substrates and refilled these in an $AsCl_3$, Ga, H_2 system. He found exceptionally good flat surfaces on these refilled holes for all substrate orientations. On the slowest growing face, the (111)B, holes of different depths could be filled to approximately the same height and little or no overgrowth occurred. As discussed by Conrad and Haisty (1966), these two facts make these systems useful for producing planar devices.

Doping is usually carried out by introducing a volatile compound, usually the element itself or a chloride, into the system between the source and seeds. Common *p*-type dopants are elemental zinc or cadmium. *n*-Type dopants are tellurium, sulphur, H_2S, H_2Se, $SnCl_4$ and $SiCl_4$.

Although vapour-phase systems using AsH_3, PH_3, gallium and hydrogen have been used extensively for growing $GaAs_{1-x}P_x$ laser material (Tietjen and Amick, 1966), vapour-phase systems have not yet achieved any prominence for GaAs laser material. However, Pilkuhn and Rupprecht (1966) have produced complex three-layer structures in GaAs by a combination of vapour-phase epitaxy and diffusion. They have produced and studied spontaneous and stimulated emission in n^+ip^+, n^+pp^+ and n^+np^+ structures.

One disadvantage in using the $AsCl_3$, Ga, H_2 system particularly for epitaxial growth on (100) orientated substrates has been the occurrence of pyramids or hillocks on otherwise flat surfaces. Joyce and Mullin (1967) suggested that these grow by a vapour–liquid–solid mechanism used by Wagner and Ellis (1964) to explain the whisker growth in silicon epitaxy. Work in this laboratory (Burling and Lockwood, 1967) showed that these defects can be completely eliminated by deliberately misorientating the substrates a few degrees off their (100) plane. Fig. 4.18 shows the growth on a (100) slice angle lapped at 3° over half its face prior to growth, the photograph clearly shows the effect of the growth plane orientation on the number of defects produced. Misorientation of more than 5° produced a heavily stepped growth.

Any dislocations present in the substrate propagate into the epitaxial layer and, in addition, any inclusions at the interface will create further defects. Abrahams and Buiocchi (1965) produced an etch which permitted the study of defects and stacking faults on (100) faces of epitaxial layers. Williams (1967) described the effect of small changes in doping levels in introducing additional defects in epitaxial layers.

FIG. 4.18. Effect of substrate orientation on epitaxial growth defects. The (100) face shows typical pyramids whilst the face 3° off (100) is free of these defects

Successful epitaxial growth directly onto single crystal metal substrates has been reported by Knappett and Owen (1967) who used the $AsCl_3$, Ga, H_2 system to produce continuous layers 15 to 20 μm thick on mechanically polished tungsten. This could be used in future devices, growing complete structures epitaxially onto metal heat-sinks.

4.6   Chemical etchants for GaAs

Etches are used for GaAs for three main reasons, the first is to remove damage after a cutting or lapping stage to produce a clean surface, the second is to reveal imperfections or crystal structure and the third is for device fabrication. It should be noted that surfaces produced after etching are not perfectly clean as they still contain contaminants adsorbed from the etchant and rinsing water. In view of this it is important to use only pure chemicals and water in the etching and rinsing steps. However, this lack of absolute cleanliness is compensated by the ease and reproducibility of the etching process for preparing GaAs surfaces.

Etchants to remove damage are usually rapid and non-selective, they produce a flat polished surface on any orientation from a lapped surface. Polishing etches are used to prepare substrates prior to epitaxial growth in both liquid- and vapour-phase systems. They are also used to prepare very thin sections in slices for electron microscope examination. Most oxidizing etchants attack GaAs by forming a soluble arsenic acid and gallium salts. Sulphuric acid is therefore a useful etch and has the added advantage that it increases the viscosity of the etchant, improving the chemical polish.

Crystal imperfections are revealed by etching in selective etchants. These produce distinctive pits at points where dislocations meet the surface and are usually orientation dependent. Dislocation densities are normally measured by counting the etch pits produced on the (111)A faces of GaAs. Before etching to reveal dislocations it is essential to ensure that the surface is free from damage, otherwise pits will also be produced at mechanically induced dislocations. Surfaces are usually best prepared by a chemical or electrolytic polish and not by a mechanical polish. Abrahams and Ekstrom (1960) discussed the use of the Schell etchant (1957) for assessing defects in GaAs. They showed a 1 : 1 correlation between the pits produced on the (111)A face and the presence of dislocations. Fig. 4.19 shows photographs of typical etch pits on the (111)A face of GaAs produced by using the Schell etchant at 80°C for 15 sec. Surface damage from the slice was removed before this stage by etching for 2 min in $3 : 1 : 1$ ($H_2SO_4 : H_2O_2 : H_2O$) etchant. Counting these pits gives a dislocation density of 10^3 cm^{-3} which is typical of good boat-grown ingots.

Selective etchants for orientations other than (111)A are described by Nasledov and coworkers (1958) and Richards and Crocker (1960). Abrahams and Buiocchi (1965) were able to reveal defects on (111)A, (111)B, (110) and (100) faces, using their etchant they also delineated stacking faults on (100) planes of epitaxial layers.

Table 4.2 summarizes some of the more useful etchants for GaAs. The conditions given in the table are not invariable and can depend on the size of the sample, agitation and starting surface. Small variations in the composition of the etchants are not usually critical to the results.

Fig. 4.19. Etch pits on the (111)A face of GaAs. The specimen was etched for 15 sec at 80°C in $HNO_3 : H_2O$ in the ratio 1 : 2
(a) Distribution of pits over the whole slice
(b) Central area of slice at higher magnification
(c) Enlargement of etch pits showing internal structure

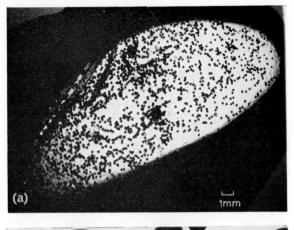

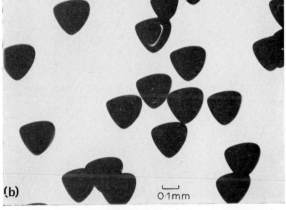

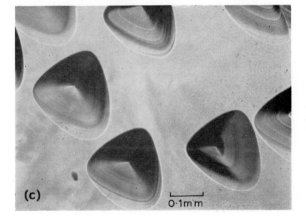

TABLE 4.2. Chemical etchants for GaAs.

Etchant and composition[a]	Conditions and remarks	Reference
$H_2SO_4 : H_2O_2 : H_2O$ 3 : 1 : 1	Polishing etch 2 min at 80°C removes $\sim 30~\mu m$ (100)	Cunnell, Edmond and Harding (1960)
$HNO_3 : HF$ 3 : 1	Polish	Richards (1960)
$HNO_3 : HF : H_2O$ 3 : 1 : 2	Polish both (111) A and B faces	Richards and Crocker (1960)
$Br_2 : CH_3OH$ 5–20% solution	Rapid polish. Used as etch for mechanical–chemical polishing	Fuller and Allison (1962)
$HNO_3 : H_2O$ 1 : 2	Etch 15 min at 20°C. Triangular pits at dislocations on A faces. Similar results at 80°C for 15 sec	Schell (1957)
$H_2O_2 : NaOH$ 1 : 5	5% NaOH solution. 5 min etch pits on A and B faces	Nasledov and others (1958)
$HNO_3 : H_2O$ ^+Ag ion 1 : 3	$\sim 1\%$ $AgNO_3$. 5 min etch pits on A and B faces	Richards and Crocker (1960)
$H_2O : AgNO_3 : CrO_3 : HF$ 2 ml : 18 g : 1 g : 1 ml	65°C for 10 min. Shows dislocations on (111) A and B, (100) and (110). Stacking faults on (100) epitaxial layers	Abrahams and Buiocchi (1965)
$HF : HNO_3 : H_2O$ 1 : 3 : 4	1 min at 15°C. Delineates nn^+, np, pp^+ junctions and shows impurity stria on cleaved (110) planes in epitaxial layers grown on (100) or (111) planes	Monsanto III–V Intermetallics Evaluation Procedures Manual (1962)

[a] Etchant compositions are by volume measurements except where stated otherwise.

4.7 Properties and distribution of impurities

4.7.1 Impurities in GaAs

Semiconductor devices usually depend on the properties of the crystalline material at a *p–n* junction. Homogeneous doping is achieved by adding a suitable impurity to the material whilst it is being grown, whereas a *p–n* junction is usually formed by diffusing an impurity of the opposite type into a homogeneously doped solid.

When GaAs is grown from a melt the distribution of impurities in the solid crystal depends on a distribution coefficient which is defined as the ratio of the concentrations on either side of the solid–liquid interface. Some experimentally determined distribution coefficients obtained from melt-growth systems are reviewed in Table 4.3. Many of the differences in the values result from different temperatures, vapour pressures and stoichiometry used in the growth of the material on which the determinations were made.

The simplest case of diffusion occurs when the diffusion coefficient is a constant at a given temperature and Fick's law is obeyed. This unfortunately is often not the case in GaAs since the rate of diffusion can depend on the arsenic pressure in the system and also on the concentration of the diffusant. Some experimentally determined diffusion coefficients are given in Table 4.4. However, these values should be treated somewhat cautiously and reference made to the original work for the specified experimental conditions.

A great number of elements are electrically active impurities in GaAs and produce shallow donor or acceptor levels. In addition a considerable number of deep levels due to impurities or lattice defects have also been reported. A summary of some of the observed energy levels is given in Table 4.5.

4.7.2 Group II elements

The most common impurities used to dope GaAs from this group are zinc and cadmium. Both elements substitute for the trivalent gallium atoms and lack by one electron those necessary to complete the tetrahedral bonds and therefore act as acceptors. They are both very soluble in GaAs, zinc reaching solubilities $> 10^{20}$ cm^{-3} at the melting point.

The diffusion of zinc in GaAs has been studied in some detail by Cunnell and Gooch (1960), who showed that the diffusion does not obey Fick's law. Weisberg and Blanc (1963) have shown that the diffusion process consists of a rapid diffusion of a low concentration of interstitial zinc combined with a slower diffusion of zinc in the gallium sub-lattice. The observed diffusion coefficient depends on the zinc concentration in the system as well as the arsenic vapour pressure.

TABLE 4.3. Distribution coefficients of common impurities in GaAs.[a]

Doping element	1	2	3	4	5
Cd		<0.02	<0.2		
Cu			2×10^{-3}		$<2 \times 10^{-3}$
Cr				6.4×10^{-4}	
Fe		<0.02	3×10^{-3}	2.0×10^{-3}	1×10^{-3}
Ni		<0.02		6.0×10^{-4}	4×10^{-5}
P		2			3
S	0.17	0.3	0.5–1.0		0.30
Se			0.44–0.55		0.30
Si	0.11	0.1	0.14		0.14
Sn	0.048	0.03			0.08
Te	0.025	0.3	0.054–0.16		0.059
Zn	0.36	0.1	0.27–0.9		0.40

[a] Values given are taken from the following:
1. Edmond (1959).
2. Weisberg, Rosi and Herkart (1960).
3. Whelan, Struthers and Ditzenberger (1960).
4. Haisty and Cronin (1964).
5. Willardson and Allred (1967).

TABLE 4.4. Diffusion coefficients in GaAs.

Element	$D = D_0 \exp(-\Delta E/kT)$		Reference
	D_0 (cm²/sec)	ΔE (eV)	
Zn	Non-Fickian		Cunnell and Gooch (1960)
Cd	0.05	2.4	Goldstein (1960)
Sn	6×10^{-4}	2.5	Goldstein and Keller (1961)
Si	Non-Fickian		Antell (1965)
S	3.7×10^3	4.0	Goldstein (1961)
Se	2.8×10^3	4.1	Goldstein (1961)
Ni	0.53	1.0	Fuller and Wolfstirn (1962)
Cu	0.03	0.52	Hall and Racette (1962)
Ag	4×10^{-4}	0.8	Boltaks and Shishijanu (1963)
Au	10^{-3}	1.0	Sokolov and Shishijanu (1964)
Mn	0.65	2.49	Seltzer (1965)

Cunnell and Gooch (1960) also reported anomalies in the diffusion of cadmium into GaAs.

TABLE 4.5. Impurity levels for various elements in GaAs at 300°K.

Element	Level (eV)	Reference
Shallow acceptors		
Zn	0·024	Sze and Irvin (1968)
Cd	0·021	Sze and Irvin (1968)
Li	0·023	Sze and Irvin (1968)
Ge	0·08	Sze and Irvin (1968)
Mg	0·012	Sze and Irvin (1968)
Shallow donors		
S	0·006	Sze and Irvin (1968)
Se	0·006	Sze and Irvin (1968)
Te	0·003	Sze and Irvin (1968)
Si	0·002	Sze and Irvin (1968)
Sn	Shallow	Sze and Irvin (1968)
Ge	Shallow	Sze and Irvin (1968)
Deep acceptors		
Cu	0·15, 0·44	Allison and Fuller (1965)
O	0·63	Sze and Irvin (1968)
Mn	0·094	Haisty and Cronin (1964)
Cr	0·79	Haisty and Cronin (1964)
Fe	0·52	Haisty and Cronin (1964)
Co	0·16	Haisty and Cronin (1964)
Ni	0·21	Haisty and Cronin (1964)

4.7.3   Group IV elements

The substitution of a group IV element does not follow the simple rules of groups II and VI which replace only one sub-lattice at a time. There are several possibilities for the substitution, they can occupy one sub-lattice only and act as a donor or acceptor, they can occupy two neighbouring sites as pairs and remain electrically neutral or they may substitute both sub-lattices in unequal numbers and can be either donors or acceptors. The most important group IV elements are silicon and tin which in normal growth both act as donors. However, at high silicon concentrations only one silicon atom in three is an electrically active donor, the other two atoms substitute on both sites and therefore compensate one another (Willardson and Allred, 1967). In addition in certain circumstances it is possible to prevent this normal substitution and produce *p*-type GaAs.

Rupprecht (1967) showed that in solution growth from gallium a low concentration of arsenic can cause silicon substitution to occur predominantly on the arsenic sub-lattice resulting in *p*-type material.

The diffusion of elements from this group into GaAs is diffiicult and requires rather specialized techniques. Goldstein and Keller (1961) were able to diffuse tin from a layer of tin plated onto the surface of the material and Van Muerch (1965) used a tin-doped layer of SiO_2 on the surface.

Antell (1965) has reported that the diffusion of silicon in GaAs does not obey Fick's law and Vieland and Seidel (1962) reported that germanium had amphoteric properties.

4.7.4 Group VI elements

Sulphur, selenium and tellurium are the most commonly used donor impurities in GaAs. These group VI elements substitute for the pentavalent arsenic atoms and, having one excess electron, behave as donors. These elements are soluble to about 10^{19} atoms/cm³, but in heavily doped material there is not a linear relationship between the number of atoms present in the system and the number assessed electrically. In this non-linear range the impurities are incorporated into the lattice as complexes (Willardson and Allred, 1967).

Attempts to diffuse the group VI elements from the vapour phase frequently result in the formation of a range of compounds on the surface of the GaAs. However, Frieser (1965) has shown that this can be overcome using compounds such as Al_2S_3 deposited on the GaAs surface as the diffusion source.

4.7.5 Other elements

Copper is generally an acceptor impurity in GaAs replacing atoms on the gallium sub-lattice. Some of the observed levels (0·14 eV, 0·44 eV) are due to copper substituted at gallium sites while other levels (0·2 eV, 0·17 eV) have been ascribed to complexes produced when copper is introduced into the lattice. The presence of copper in GaAs is generally believed to degrade devices and great care is necessary to avoid contamination in producing and handling the material. It is also an extremely rapidly diffusing impurity and can readily convert *n*-type GaAs into *p*-type material (Fuller, 1967). The electrical activity and solubility of copper has been studied by Blanc and Weisberg (1964) who showed the possibility of its removal by treating the GaAs with potassium cyanide.

Oxygen can be incorporated in GaAs and results in an energy level (0·63 eV) close to the centre of the band gap. Diffusion of oxygen into *n*-type GaAs produces semi-insulating compensated material with a low carrier concentration. Haisty, Mehal and Stratton (1962) have discussed in some detail the properties of oxygen-doped GaAs.

Several transition elements are used as impurities in GaAs. The acceptor properties of the elements chromium, manganese, iron, cobalt and nickel have been discussed by Haisty and Cronin (1964). A range of deep levels are observed ranging from 0·1 eV for manganese to 0·76 eV for chromium. Cronin and Haisty (1964) described in detail the production of semi-insulating ($> 10^6 \Omega$ cm) GaAs by chromium doping GaAs in a Czochralski puller. The properties of iron-, cobalt- and nickel-doped electroluminescent diodes grown by liquid epitaxy have been described by Strack (1967).

Little work on the diffusion of these transition elements in GaAs has been reported. However, Seltzer (1965) studied the diffusion of manganese as a function of both temperature and arsenic pressure. In some cases a simple error function impurity profile was obtained, but in other cases the profile was more complex and was best described by the superposition of two error function curves.

4.8 Electrical properties

4.8.1 Band structure

[i] *Pure GaAs.* A knowledge of the band structure of GaAs is essential in understanding many of the physical and optical properties of the material. Early work by Herman (1955) and Callaway (1957) deduced the GaAs energy band structure by extending related calculations on germanium and silicon. Ehrenreich (1960) reviewed the available evidence on the band structure and concluded that the material has a direct band gap with the lowest minimum in the conduction band at the centre of the zone with a second minima along the [100] directions, 0·36 eV higher. He also showed that the valence-band structure is similar to germanium with a heavy-hole band, a light-hole band and a band with holes of intermediate mass split off from the other two bands by spin–orbit interaction.

From the evidence available at the time Ehrenreich showed that the energy gap E_G and the separation of the conduction band minima ΔE could be expressed by the following equations

$$E_G = 1\cdot53 + 9\cdot4 \times 10^{-6} P - 4\cdot9 \times 10^{-4} T$$
$$\Delta E = 0\cdot36 - 1\cdot08 \times 10^{-5} P - OT$$

When P is the pressure in atmospheres and T the temperature in °K the expression gives the energy gap in eV.

The linear dependence of E_G with temperature is only valid above 100°K and more accurate values of the energy gap are given by the optical absorption measurements of Sturge (1962). These values, corrected for exciton absorption, are given below

$T(^\circ K)$	21	55	90	185	294
E_G (eV)	1·521	1·518	1·511	1·479	1·435

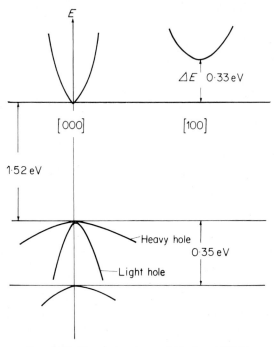

FIG. 4.20. Band structure of GaAs at 77°K

It can be seen that the linear formula given by Ehrenreich gives somewhat lower values than Sturge's values except at the low-temperature extrapolation.

The exact value of the energy separation ΔE between [000] and [100] conduction band minima has been the subject of some discussion in recent years as it is of considerable importance in Gunn devices and in the band structure of gallium arsenide–phosphide alloys. In a review of the available evidence Hilsum (1966) concludes that at 77°K the separation is $\sim 0·33$ eV and increases somewhat with temperature. The probable band structure, at 77°K, is sketched in Fig. 4.20.

[ii] *Impure GaAs.* At low concentrations the shallow donors or acceptors form discrete energy levels with finite but small ionization

energies. However, as Pankove (1965) showed, in his work on the absorption edge of doped GaAs, at high donor or acceptor concentrations the simple band structure is considerably perturbed. As the impurity concentration is increased the impurity levels form an impurity band which ultimately merges with the valence or conduction band edge. This perturbs the parabolic distribution of states and results in tails extending into the energy gap. The existence of this tail is of considerable importance in the operation of lasers at temperatures near 300°K.

According to Pankove, the distribution of these tailed states varies as $\exp E/E_0$ where E_0 is a parameter which depends on the impurity concentration. The effect becomes significant at lower donor than acceptor concentrations. Stern (1966), in extensive calculations on the effect of band tails on the properties of lasers, has suggested that the impurity level merges with the adjacent band edge for donor concentrations above 10^{17} cm^{-3} and for acceptor concentrations above 3×10^{18} cm^{-3}.

[iii] *Mixed III–V compounds.* The band structure of certain mixed crystals of GaAs and other III–V compounds is of considerable interest because it is possible to make lasers from those systems which have direct band structures.

One case of particular interest is the GaAs–GaP system since this system can provide lasers emitting in the visible part of the spectrum. As already mentioned, GaAs has a direct band gap whereas that of GaP is indirect with a band gap of 2·32 eV at 77°K and the centre conduction band minimum some 0·56 eV higher. As shown in Fig. 4.21, the main features of the band structure of the GaAs$_x$P$_{1-x}$ mixed crystals can be produced by a linear interpolation between the structures of the two pure materials.

It can be seen that material with a composition between pure GaAs and GaAs$_{0·6}$P$_{0·4}$ has a direct band gap structure with the direct band gap increasing to a maximum of 2 eV as the proportion of phosphorus is increased. Lasers have been produced from this system extending the wavelength range from 8500 to 6500°Å at 77°K.

Other mixed systems are also likely to be interesting, particularly the GaAs–AlAs system, which would have a direct band gap of nearly 2·0 eV for a crystal of Ga$_{0·5}$Al$_{0·5}$As (Ku and Black, 1966).

4.8.2 *Effective mass*

[i] *Electron effective mass.* The effective mass of an electron in the conduction band of a semiconductor can be determined from several experiments including measurements of Faraday rotation, reflectivity and infrared cyclotron resonance absorption. Early measurements on GaAs by

Moss and Walton (1959), Spitzer and Whelan (1959), Cardona (1961) and
Palik, Stevenson and Wallis (1961) gave values of $m^*/m \sim 0.072$. More
recent determinations by Ukhanov and Mal'tsev (1963) and Piller (1966)

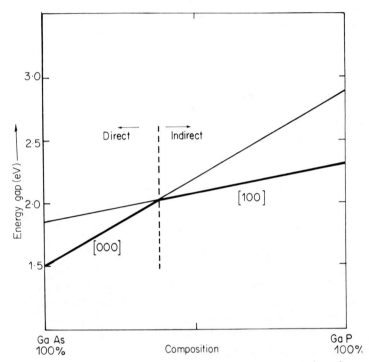

FIG. 4.21. Energy gap for the GaAs–GaP system plotted
against percentage composition

gave significantly lower values. Piller's results show the variation of
effective mass with carrier concentration, the values obtained being:

Carrier concentrations (cm^{-3})	7×10^{18}	5×10^{17}	3×10^{16}
Effective mass	0·092	0·078	0·071

and $m^*/m_0 = 0.066 \pm 0.002$. This is in good agreement with the value of
0·068 from the Russian work.

[ii] *Hole effective mass.* The hole effective masses are less well established. Braunstein (1959) showed that the heavy-hole mass is six times the
light-hole mass from his measurements on free hole absorption. Edmond,
Broom and Cunnell (1956) reported a value of m_h^* of $0.5m_0$ for the heavy-
hole mass from thermoelectric measurements. An upper limit can be

deduced from the observed hole mobility and this suggests a value of approximately $0.6m_0$.

The probable values for the electron and hole effective masses are then

$$m_e^* = 0.066m_0$$
$$m_h^* = 0.6m_0$$

4.8.3   Carrier mobility

The ionic component of the crystal binding in GaAs results in a strong carrier interaction with the optical mode lattice vibrations. The carrier mobility in pure material will then be limited by polar scattering. Hilsum (1960) calculated values for the polar mobility in the III–V compounds and obtained the following values:

Electron mobility at 300°K $\mu_e = 10,400$ cm²/V sec

Hole mobility at 300°K $\mu_h = 520$ cm²/V sec

Since acoustic scattering gives a mobility which is proportional to $(m^*)^{\frac{5}{2}}$ relatively small changes in the value of the effective mass have a significant effect on the predicted mobility. The more recent revision of the electron effective mass values which have been noted will make these values approximately 10% higher. Also, since the hole effective mass is not known with any great certainty, the predicted hole mobility is only an approximate value:

Most semiconductor devices, in particular lasers, use fairly heavily doped material and, in this case, the carrier mobilities will be determined by the combination of polar scattering and ionized impurity scattering. At these higher carrier concentrations we might also expect carrier–carrier scattering to be significant and, particularly in compensated material, scattering by neutral impurity centres may be important. Unfortunately, in the case of polar scattering, it is not possible to define a relaxation time so that the combination of scattering processes cannot be accomplished by the simple technique of combining relaxation times. However, Ehrenreich (1960) has shown how one may combine polar and ionized impurity scattering using a variational technique to obtain carrier mobilities as a function of ionized impurity concentration and temperature.

The room temperature electron mobility of a range of *n*-type samples is shown in Fig. 4.22 together with the theoretical mobility calculated by Ehrenreich (1960). It should be noted that this theoretical curve should be revised to take into account the more recent estimates of the lattice mobility.

The experimental points shown in Fig. 4.22 were obtained from samples of GaAs made by several techniques at this laboratory and also from some published data by Ainslie, Blum and Woods (1962), Oliver (1962), Whitaker and Bolger (1966) and Knight, Effer and Evans (1965).

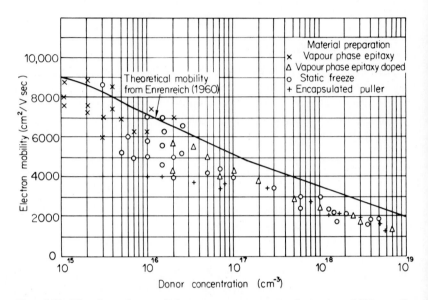

FIG. 4.22. The dependence of the room temperature electron mobility on the donor concentration

As might be expected the observed mobilities in general lie below the calculated values due to residual impurities and scattering centres in the material. The purest material is obtained using low temperature epitaxial techniques although high mobility GaAs can be boat grown using an oxygen atmosphere to suppress the silicon contamination. Table 4.6 shows some of the highest reported mobilities for the various growth methods.

The room temperature hole mobility for a range of p-type samples is shown in Fig. 4.23. For hole concentrations up to 10^{19} cm^{-3} the results were obtained from Hall coefficient and resistivity measurements on homogeneous samples by Emel'yanenko and coworkers (1960) and Rosi and coworkers (1960). The results for carrier concentrations between 10^{19} and 10^{20} cm^{-3} were determined by measurements on diffused samples (Gooch, 1965), which show values as low as 50 cm^2/V sec for a carrier concentration of 10^{20} cm^{-3}.

TABLE 4.6. Pure GaAs from several systems showing
mobilities at 300 and 77°K.

Method of preparation	Carrier concentration (cm^{-3})	Mobility (cm^2/V sec) 300°K	Mobility (cm^2/V sec) 77°K	Reference
Bridgman	3×10^{15}	8650	30,000	Ainslie, Blum and Woods (1962)
Vapour phase	2×10^{15}	8800	38,000	Effer (1965)
	7×10^{14}	8600	106,000	Bolger and coworkers (1965)
Liquid phase	2×10^{14}	7900	52,000	Goodwin and co-workers (1968)
	1×10^{14}	9300	95,000	Kang and Greene (1967)
	—	8500	106,000	Andre and Le Duc (1968)

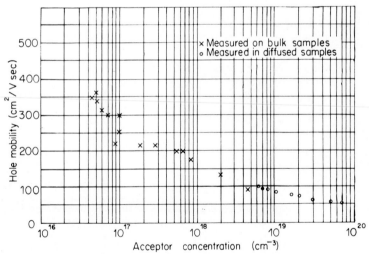

FIG. 4.23. The dependence of the room temperature hole
mobility on the acceptor concentration

The temperature dependence of electron and hole mobilities for samples
with different carrier concentrations is shown in Fig. 4.24. Data for this

figure were obtained from Edmond, Broom and Cunnell (1956), Weisberg Rosi and Herkart (1960), Rosi, Meyerhofer and Jensen (1960), Oliver (1962) and Kang and Greene (1967).

For heavily doped material, such as used for semiconductor lasers, the mobility is relatively insensitive to temperature. For donor concentrations

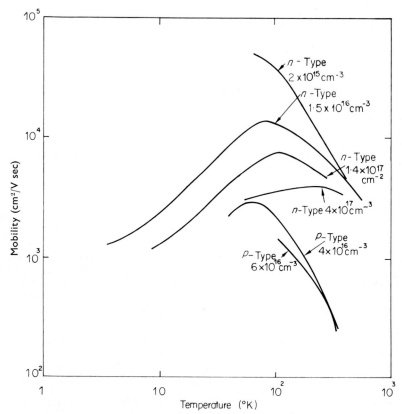

Fig. 4.24. Temperature dependence of electron and hole mobilities for samples of different carrier concentrations

about 2×10^{18} cm^{-3} it is sufficiently accurate for most purposes to take a value of 2000 cm^2/V sec over the temperature range 77 to 300°K.

In pure material the mobility rises to a maximum value at 80°K and for this reason measurements of mobility at this temperature give a convenient means of assessing the purity of the material. Mobilities measured at this temperature are included in Table 4.6.

4.8.4 Hall coefficient and electrical conductivity

The Hall coefficient R_H and electrical conductivity σ are measurable material properties which depend on the impurity carrier concentration and mobility.

$$R_H = 1/rne \tag{4.14}$$

$$\sigma = ne\mu \tag{4.15}$$

In Eqn. (4.14) the numerical coefficient r is included to take account of the scattering mechanism and is different for each mechanism. In practice r is taken as unity and the carrier mobility is deduced from measurements of R_H and σ. Hence quoted values of mobilities $\mu = R_H \sigma$ are usually Hall mobilities.

n-Type samples have very low activation energies as shown in Table 4.4. The purest available *n*-type GaAs has an electron concentration about 10^{15} cm^{-3} whereas the intrinsic concentration is approximately 10^6 cm^{-3} at room temperature. Hence intrinsic conductivity is only observed at high temperatures and in this region an appreciable number of electrons populate the [100] conduction band with a lower mobility.

At low temperatures pure samples show impurity band conduction in a donor band 0·01 eV below the conduction band. The mobility in this band increases as the donor concentration increases (Emel'yanenko and Nasledov, 1958; and Oliver, 1962). In samples with impurity concentrations above 10^{17} cm^{-3} the impurity band overlaps the conduction band and the carrier concentration is essentially independent of temperature over the temperature range 4 to 500°K.

The activation energy of the acceptor levels is about 0·02 eV and can be determined from accurate measurements of the Hall coefficient. In this case the impurity levels do not overlap the conduction band below impurity concentrations of 3×10^{18} cm^{-3}. *p*-Type material can, therefore, show impurity band conduction at low temperatures and the carrier concentration is temperature dependent up to relatively high acceptor concentrations.

4.8.5 Semi-insulating GaAs

In the discussion on the electrical properties of GaAs we have assumed the material to be semiconducting with a room-temperature resistivity in the range 0·1 to 10^{-3} Ω cm. Such material is essentially uncompensated and the carrier concentration is determined by the donor or acceptor concentration. It has not been possible to produce pure material with an uncompensated carrier concentration less than 10^{15} cm^{-3} but it is possible to produce compensated GaAs with a carrier concentration approaching the intrinsic value ($\sim 10^6\,\text{cm}^{-3}$ at room temperature). This material will have a resistivity between 10^6 and $10^9\,\Omega\,\text{cm}$ at room temperature and is known as high resistivity or, more commonly, semi-insulating GaAs.

Semi-insulating GaAs is not a particularly pure material, but contains a large number of impurities with deep levels, which compensate the normally present donors and acceptors.

Two models have been proposed to explain the properties of semi-insulating GaAs.

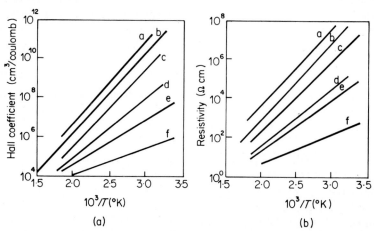

FIG. 4.25. The temperature dependence of the Hall coefficient (a) and resistivity (b) of semi-insulating GaAs

The first model by Allen (1960) proposed that the residual *n*-type impurity was compensated by a deep acceptor which could be oxygen or chromium. These acceptor impurities could have two positions in the lattice, one electrically neutral, the other a deep acceptor. Allen suggested an automatic compensating mechanism in which just sufficient impurities were electrically active to compensate the residual donors present.

A second model proposed by Blanc and Weisberg (1961) suggested the existence of a deep donor level and a shallow acceptor level to compensate the residual donors. This model does not require the self-compensating mechanism and high resistivity GaAs results independent of the exact concentration of the deep donors.

Both these models explain many of the properties of semi-insulating GaAs but there are also objections to both models and these are covered in a detailed discussion by Hilsum (1965).

Semi-insulating GaAs results from float-zone refining and from the growth of crystals in an oxygen-containing atmosphere. The usual method to grow semi-insulating GaAs is to dope crystals with chromium, a deep acceptor. It is necessary to resort to the vapour phase to produce GaAs having a resistivity between the two extremes of semiconducting and semi-insulating material.

The Hall coefficient of some samples of high resistivity GaAs has been determined by Gooch (1962) and these are shown in Fig. 4.25. The highest resistivity samples have an activation energy of 0·76 eV which is exactly that which would be expected for intrinsic GaAs. However, this material must be compensated since the same activation energy is observed over a range of carrier concentrations and the mobility of the material is low, about 2,000 cm²/V sec (Gooch, Hilsum and Holeman, 1961).

Semi-insulating *p*-type GaAs can also be obtained, although its production is less controllable than that of *n*-type material as it appears that *p*-type material requires a very accurate control of the impurities introduced. Due to the high electron to hole mobility ratio, two carrier effects are observed at high temperatures and the sign of the Hall coefficient reverses as the intrinsic region is approached.

4.9   Other properties

4.9.1   Optical absorption

As far as semiconductor lasers are concerned the most important part of the absorption spectrum of GaAs is that in the vicinity of the band gap energy. The fundamental absorption edge can best be studied by measurements on semi-insulating GaAs which has few carriers and therefore little free-carrier absorption. Studies on the absorption spectra by Moss (1961) and Sturge (1962) show that the absorption coefficient rises extremely sharply with energy having an exponential dependence up to coefficients of approximately 8,000 cm⁻¹. Sturge (1962) studied the absorption for photon energies between 0·6 and 2·7 eV over the temperature range 10 to 300°K. These measurements probably provide the most accurate values of the band gap of GaAs and show the existence of an exciton with a binding energy of 3 meV.

The absorption edge at 90 and 290°K is shown in Fig. 4.26 from data given by Sturge. As to be expected for a direct band gap semiconductor this edge is extremely steep, rising from 1 to 10^4 cm⁻¹ for a photon energy change of 0·04 eV. Calculations from the absorption data gave an energy gap of 1·435 eV at 290°K and 1·521 eV at 21°K.

The absorption edge in impure GaAs has been studied by Turner and Reese (1964) and Pankove (1965). Fig. 4.27 shows some of the results obtained by Turner and Reese for both *n*- and *p*-type samples doped to the levels required for lasers. Curves are shown for temperatures of 77

7

and 300°K and carrier concentrations $9 \times 10^{17}\,\text{cm}^{-3}$ (*n*-type) and $2 \times 10^{19}\,\text{cm}^{-3}$ (*p*-type).

At the absorption edge the coefficient is found to increase exponentially with photon energy and also to move to higher energies as the carrier concentration is increased. Pankove (1965) has shown that these results reflect the presence of tails of states extending the valence and conduction bands into the band gap. The distribution of these tail states is exponential varying as $\exp E/E_0$ where E_0 is a parameter which increases with doping. E_0 becomes significant at lower concentrations of donors than acceptors.

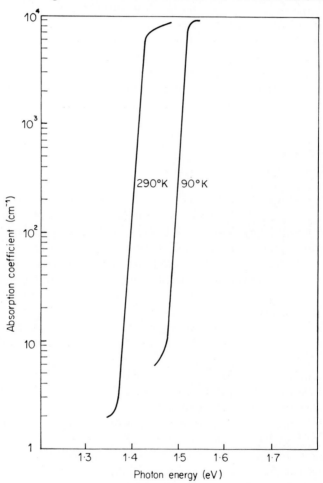

FIG. 4.26. The absorption edge of GaAs at 90 and 290°K

The shift of the absorption edge with temperature is greater at temperatures above 100°K than it is from 0 to 100°K. Using data presented by Pankove (1965) for an *n*-type sample with a carrier concentration of 6.8×10^{18} cm^{-3}, the temperature dependence from 100 to 300°K is -6.4×10^{-4} eV/°K and between 0 and 100°K it is -2.9×10^{-4} eV/°K.

4.9.2 *Refractive index and dielectric constant*

The refractive index of GaAs is dependent on wavelength, carrier concentration and temperature. Measurement at wavelengths well away

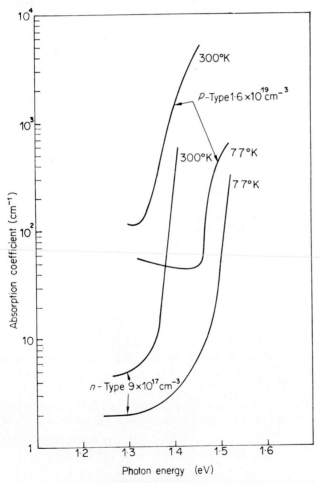

FIG. 4.27. Absorption coefficient data for doped GaAs as a function of photon energy

from the band edge (Hambleton, Hilsum and Holeman, 1961), using wavelengths below 0·65 eV, gives a refractive index of 3·347. Marple (1964) obtained refraction data for photon energies from 0·7 eV to the band edge, using an undoped prism of GaAs. Fig. 4.28 shows the refractive index as a function of photon energy at temperatures between 100 and 300°K. Taking values of 1·46 and 1·38 eV as the emission energies for lasers at 100 and 300°K the corresponding refractive indices are 3·58 and 3·60.

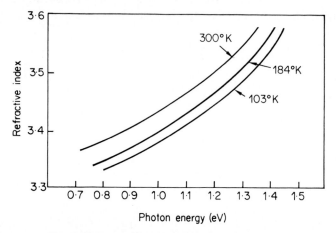

FIG. 4.28. Refractive index of GaAs as a function of photon energy (after Marple, 1964)

Hambleton, Hilsum and Holeman (1961) obtained a value of 10·90 for the high frequency dielectric constant using infrared refraction measurements.

4.9.3 Thermal conductivity

GaAs lasers are operated at high power densities so that a considerable amount of heat must be removed from the device. It is for this reason that the thermal conductivity is an important parameter for GaAs laser technology.

Holland (1964) studied thermal conductivity in GaAs as a function of temperature for *n*-type samples. He used these results to investigate the validity of relaxation time expressions for various phonon interactions and scattering effects in calculating theoretical values for the thermal conductivity.

Carlson, Slack and Silverman (1965) determined values for both *n*- and *p*-type GaAs over a temperature range of 3 to 300°K. Fig. 4.29 shows some of these results for carrier concentrations of 10^{16} to 10^{18} cm^{-3} for donors and 10^{18} to 10^{19} cm^{-3} for acceptors.

Results of thermal conductivity values obtained by many authors at temperatures above 77°K on both *p*- and *n*-type material show only small differences over the range of impurities from 10^{15} to 10^{20}. These data are

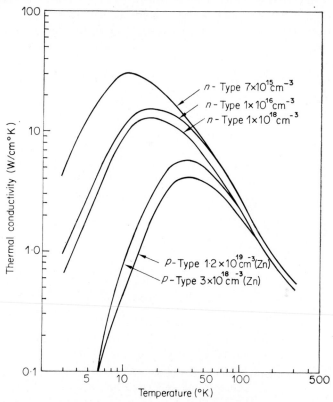

FIG. 4.29. The thermal conductivity of GaAs as a function of temperature for samples of different carrier concentrations

shown in Fig. 4.30 for temperatures of 77 and 300°K. For most practical purposes it is sufficiently accurate to ignore this variation due to the impurity concentration and take the following values:

$$\text{Thermal conductivity} \begin{cases} 77°\text{K} = 2\cdot5 \pm 0\cdot5 \text{ W/cm °K} \\ 300°\text{K} = 0\cdot45 \pm 0\cdot05 \text{ W/cm °K} \end{cases}$$

As shown in Fig. 4.29, at lower temperatures the difference between *p*- and *n*-type GaAs becomes significant, the highest conductivity being obtained with *n*-type material with a low carrier concentration. The maximum conductivity is observed at 10–20°K so that in some cases it is more advantageous to operate lasers at liquid hydrogen temperatures rather than liquid helium temperatures. In this low-temperature range the thermal conductivity of *p*-type GaAs can be as much as two orders or more lower than *n*-type GaAs, so that heat should be removed through the *n* side of the device.

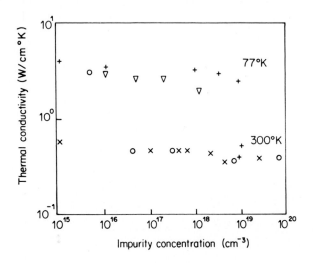

+ Carlson (1965)   o Amick, Kudman and Steigmeier (1965)
∇ Holland (1964)   × Aliev and Sholyt (1965)

FIG. 4.30. The thermal conductivity of GaAs as a function of impurity concentration at 77 and 300° K

4.9.4 *Thermal expansion and lattice constant*

The thermal expansion coefficient of GaAs has recently been determined by Pierron, Parker and McNeely (1966) using accurate x-ray measurements of lattice dimensions over the temperature range from −62 to 200°C. For this range they determined an expansion coefficient of

$$6 \cdot 86 \times 10^{-6} \pm 0 \cdot 13 \times 10^{-6} \ °C^{-1}.$$

At 24°C the lattice constant is 5·6532 Å which agrees well with Straumanis and Kim (1965) who obtained $5 \cdot 65321 \pm 3 \times 10^{-5}$ Å at 25°C.

4.9.5 Specific heat

The most extensive study of specific heat measurements on GaAs was carried out by Piesbergen (1963) over the temperature range 12 to 275°K. Fig. 4.31 shows these results plotted in W/g °K as a function of temperature.

From these results Piesbergen calculated the Debye temperature and obtained a value of $355 \pm 5°$K which compares favourably with the value of $345 \pm 5°$K deduced by Garland and Pasks (1962) from measurements of the elastic constants.

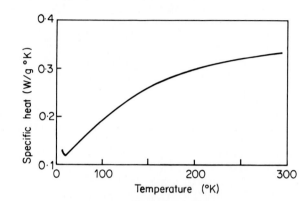

FIG. 4.31. Specific heat of GaAs as a function of temperature (after Piesbergen, 1963)

For the two temperatures of particular interest in GaAs laser work we have the following values from Piesbergen's work for the specific heat of GaAs:

$$\text{Specific heat} \begin{cases} 77°\text{K} = 0\cdot16 \text{ W/g °K} \\ 300°\text{K} = 0\cdot32 \text{ W/g °K} \end{cases}$$

4.9.6 Density

The experimental density of GaAs at the stoichiometric composition has been determined by Straumanis and Kim (1965) who obtained the value

$$5\cdot3176 \pm 0\cdot0026 \text{ g/cm}^3$$

at 25°C. The density determined from x-ray data was $5\cdot3167$ g/cm^3 at the same temperature.

4.9.7 Elastic moduli

Measurements of the adiabatic moduli of GaAs at 25°C were made by Bateman, McSkimin and Whelan (1959). The three moduli required to specify the stress–strain relationship and the values obtained are:

$$C_{11} = 1{\cdot}188 \times 10^{12} \text{ dyn/cm}^2$$

$$C_{12} = 0{\cdot}538 \times 10^{12} \text{ dyn/cm}^2$$

$$C_{44} = 0{\cdot}594 \times 10^{12} \text{ dyn/cm}^2$$

References

Abrahams, M. S., and L. Ekstrom (1960). *Properties of Elemental and Compound Semiconductors*, Ed. H. C. Gatos, Interscience Publishers, New York, pp. 225–241.
Abrahams, M. S., and C. J. Buiocchi (1965). *J. Appl. Phys.*, **36**, 2855.
Abrahams, M. S., and C. J. Buiocchi (1966). *J. Appl. Phys.*, **37**, 1973.
Ainslie, N. G., S. E. Blum and J. F. Woods (1962). *J. Appl. Phys.*, **33**, 2391.
Aliev, S. A., and S. S. Sholyt (1965). *Fiz. Tverd. Tela.*, **7**, 3690. (*Soviet Phys. Solid State*, **7**, 2986, 1966.)
Allen, J. W. (1960). *Nature*, **187**, 403.
Allison, H. W., and C. S. Fuller (1965). *J. Appl. Phys.*, **36**, 2519.
Amick, A., I. Kudman and E. F. Steigmeier (1965). *Phys. Rev.*, **138**, A1270.
Andre, E., and J. M. LeDuc (1968). *Mat. Res. Bull.*, **3**, 1.
Antell, G. R. (1965). *Solid-State Electron.*, **8**, 943.
Antell, G. R., and D. Effer (1959). *J. Electrochem. Soc.*, **106**, 509.
Arthur, J. R. (1967). *J. Phys. Chem. Solids*, **28**, 2257.
Bass, S. J., and P. E. Oliver (1967). *Proc. Intern. Sym. on GaAs, Reading 1966*, Inst. Phys. and Phys. Soc. London, pp. 41–45.
Bateman, T. B., H. J. McSkimin and J. M. Whelan (1959). *J. Appl. Phys.*, **30**, 544.
Blanc, J., and L. R. Weisberg (1961). *Nature*, **192**, 155.
Blanc, J., and L. R. Weisberg (1964). *J. Phys. Chem. Solids*, **25**, 221.
Blanc, J., R. H. Bube and L. R. Weisberg (1964). *J. Phys. Chem. Solids*, **25**, 225.
Blum, S. E. (1962). *Compound Semiconductors, Vol. 1, Preparation of III–V Compounds*, Eds. R. K. Willardson and H. L. Goering, Reinhold Publishing Corporation, New York, pp. 85–91.
Bolger, D. E., J. Franks, J. Gordon and J. Whitaker (1967). *Proc. Intern. Sym. on GaAs, Reading 1966*, Inst. Phys. and Phys. Soc. London, p. 16.
Boltaks, B. I., and F. S. Shishijanu (1963). *Fiz. Tverd. Tela.*, **5**, 2310. (*Soviet Phys. Solid State*, **5**, 1680, 1964.)
Braunstein, R. (1955). *Phys. Rev.*, **99**, 1892.
Braunstein, R. (1959). *J. Phys. Chem. Solids*, **8**, 280.
Brice, J. C. (1965). *The Growth of Crystals from the Melt*, North-Holland Publishing Co., p. 127.
Brice, J. C. (1967). *Solid State Electron.*, **10**, 335.
Bridgman, P. W. (1925). *Proc. Am. Acad. Arts. Sci.*, **60**, 305.
Broom, R. F. (1963). *Phys. Letters*, **4**, 330.
Burling, M., and K. J. Lockwood (1967). Services Electronics Research Laboratory, private communication.
Calloway, J. (1957). *J. Electron.*, **2**, 230.
Cardona, M. (1961). *Phys. Rev.*, **121**, 752.

Carlson, R. O. (1965). *J. Appl. Phys.*, **36**, 505.
Carlson, R. O., G. A. Slack and S. J. Silverman (1965). *J. Appl. Phys.*, **36**, 505.
Cochran, C. N., and L. M. Foster (1962). *J. Electrochem. Soc.*, **109**, 149.
Conrad, R. W., and R. W. Haisty (1966). *Trans. Met. Soc. AIME*, **236**, 263.
Cronin, G. R., M. E. Jones and O. Wilson (1963). *J. Electrochem. Soc.*, **110**, 582.
Cronin, G. R., and R. W. Haisty (1964). *J. Electrochem. Soc.*, **111**, 874.
Cunnell, F. A., J. T. Edmond and W. R. Harding (1960). *Solid-State Electron.*, **1**, 97.
Cunnell, F. A., and C. H. Gooch (1960). *J. Phys. Chem. Solids*, **15**, 127.
Cunnell, F. A., and R. Wickham (1960). *J. Sci. Instr.*, **37**, 410.
Czochralski, J. (1917). *Z. Phys. Chem.*, **92**, 219.
De La Breteque, P. (1962). *Compound Semiconductors, Vol. 1, Preparation of III–V Compounds*, Eds. R. K. Willardson and H. L. Goering, Reinhold Publishing Corporation, New York, pp. 68–73.
Dousmanis, G. C., H. Nelson and D. L. Staebler (1964). *Appl. Phys. Letters*, **5**, 174.
Edmond, J. T. (1959). *Proc. Phys. Soc.*, **73**, 622.
Edmond, J. T., R. F. Broom and F. A. Cunnell (1956). *Semiconductor Meeting Report*, Rugby: *London Phys. Soc.*, p. 109.
Effer, D. (1965). *J. Electrochem. Soc.*, **112**, 1020.
Ehrenreich, H. (1960). *Phys. Rev.*, **120**, 1951.
Ekstrom, L., and L. R. Weisberg (1962). *J. Electrochem. Soc.*, **109**, 321.
Emel'yanenko, O. V., and D. N. Nasledov (1958). *Zh. Tekh. Fiz.*, **28**, 1177. (*Soviet Phys.-Tech. Phys.*, **3**, 1094, 1958.)
Emel'yanenko, O. V., T. S. Lagonova and O. N. Nasledov (1960). *Fiz. Tverd. Tela.*, **2**, 192. (*Soviet Phys. Solid State*, **2**, 176.)
Folberth, O. G. (1962). *Compound Semiconductors, Vol. 1, Preparation of III–V Compounds*, Eds. R. K. Willardson and H. L. Goering, Reinhold Publishing Corporation, New York, pp. 21–33.
Freiser, R. G. (1965). *J. Electrochem Soc.*, **112**, 697.
Fuller, C. S. (1967). *J. Appl. Phys.*, **38**, 2873.
Fuller, C. S., and H. W. Allison (1962). *J. Electrochem. Soc.*, **109**, 880.
Fuller, C. S., and K. B. Wolfstirn (1962). *J. Appl. Phys.*, **33**, 2507.
Fuller, C. S., K. B. Wolfstirn and H. W. Allison (1967). *J. Appl. Phys.*, **38**, 4339.
Garland, C. W., and K. C. Pasks (1962). *J. Appl. Phys.*, **33**, 759.
Gates, H. C., and M. C. Lavine (1960). *J. Electrochem. Soc.*, **107**, 427.
Goldschmidt, V. M. (1929). *Trans. Faraday. Soc*, **25**, 253.
Goldsmith, N. (1963). *J. Electrochem. Soc.*, **110**, 588.
Goldstein, B. (1960). *Phys. Rev.*, **118**, 1024.
Goldstein, B. (1961). *Phys. Rev.*, **121**, 1305.
Goldstein, B., and H. Keller (1961). *J. Appl. Phys.*, **32**, 1180.
Gooch, C. H. (1962). Services Electronics Research Laboratory, private communication.
Gooch, C. H. (1965). *Phys. Letters*, **3**, 183.
Gooch, C. H., C. Hilsum and B. R. Holeman (1961). *J. Appl. Phys.*, **32**, 2069.
Goodwin, A. R., J. Gordon and C. D. Dobson (1968). *Brit. J. Appl. Phys.*, **1**, 115.

Gottlieb, G. E., and J. F. Corboy (1963). *RCA Rev.*, **24**, 585.

Gremmelmaier, R. (1956). *Z. Naturforsch.*, **11a**, 511.

Haisty, R. W., and G. R. Cronin (1964). *Proc. 7th Int. Conf. on the Physics of Semiconductors*, Paris, Dunod; New York, Academic Press, p. 1161.

Haisty, R. W., E. W. Mehal and R. Stratton (1962). *J. Phys. Chem. Solids*, **23**, 829.

Hall, R. N. (1963). *J. Electrochem. Soc.*, **110**, 385.

Hall, R. N., G. E. Fenner, J. D. Kingsley, T. J. Soltys and R. O. Carlson (1962). *Phys. Rev. Letters*, **9**, 366.

Hall, R. N., and J. H. Racette (1962). *Bull. Am. Phys. Soc.*, **7**, 234.

Hambleton, K. G., C. Hilsum and B. R. Holeman (1961). *Proc. Phys. Soc. London*, **77**, 1147.

Hara, T., and I. Akasaki (1966). *Japan J. Appl. Phys.*, **5**, 1255.

Herman, F. (1955). *J. Electron.*, **1**, 103.

Hilsum, C. (1960). *Proc. Phys. Soc. London*, **76**, 414.

Hilsum, C. (1965). *Progress in Semiconductors*, Vol. 9, Eds. A. F. Gibson and R. E. Burgess, Temple Press Books Ltd., London, p. 137.

Hilsum, C. (1965). *Progr. Semiconductors*, **9**, 154.

➤ Hilsum, C. (1966). *Semiconductors and Semimetals, Vol. 1*, Eds. R. K. Willardson and A. C. Beer, Academic Press, New York, p. 4.

Holland, M. G. (1964). *Phys. Rev.*, **134**, A471.

Hurle, D. T. J. (1962). *Progr. Mat. Sci.*, **10**, 79.

Joyce, B. D., and J. B. Mullin (1967). *Proc. Intern. Sym. on GaAs, Reading, 1966*, Inst. Phys. and Phys. Soc. London, pp. 23–26.

Kang, C. S., and I. E. Greene (1967). *Appl. Phys. Letters*, **11**, 171.

Keyes, R. J., and T. M. Quist (1962). *Proc. IRE*, **50**, 1822.

Knappett, J. E., and S. J. T. Owen (1967). *Phys. Stat. Sol.*, **21**, K99.

Knight, J. R. (1961). *Nature*, **190**, 1001.

Knight, J. R., D. Effer and P. R. Evans (1965). *Solid State Electron.*, **8**, 178.

Koster, W., and B. Thoma (1955). *Z. Metallk.*, **46**, 291.

Ku, S. M., and J. F. Black (1966). *J. Appl. Phys.*, **37**, 3733.

Madelung, O. (1964). *Physics of III–V Compounds*, John Wiley and Sons, New York, p. 104.

Marple, D. T. F. (1964). *J. Appl. Phys.*, **35**, 1241.

McCarthy, J. P. (1966). *J. Appl. Phys.*, **38**, 436.

Metz, E. P. A., R. C. Miller and R. Mazelsky (1962). *J. Appl. Phys.*, **33**, 2016.

Moody, P. L., and C. Kolm (1958). *Rev. Sci. Instr.*, **29**, 1144.

Moss, T. S. (1961). *J. Appl. Phys.*, **32**, 2136.

Moss, T. S., and A. K. Walton (1959). *Proc. Phys. Soc.*, **74**, 131.

Mlavsky, A. I., and M. Weinstein (1963). *J. Appl. Phys.*, **34**, 2885.

Mullin, J. B., B. W. Straughan and W. S. Brickell (1965). *J. Phys. Chem. Solids*, **26**, 782.

Nasledov, D. N., A. Ya, Patrakova and B. V. Tsarenkov (1958). *Zh. Tekh. Fiz.*, **28**, 779. (*Soviet Phys.-Tech. Phys.*, **3**, 726, 1958.)

Nathan, M. I., W. P. Dumke, G. Burns, F. H. Dill, Jr. and G. Lasher (1962). *Appl. Phys. Letters*, **1**, 62.

Nelson, H. (1963). *RCA Rev.*, **24**, 603.

Nelson, H., J. I. Pankove, F. Hawrylo, G. C. Dousmanis and C. Reno (1964). *Proc. I.E.E.E.*, **52**, 1360.

Nicoll, F. H. (1963). *J. Electrochem. Soc.*, **110**, 1165.

Oliver, D. J. (1962). *Phys. Rev.*, **127**, 1045.

Palik, E. D., J. R. Stevenson and R. F. Wallis (1961). *Phys. Rev.*, **124**, 701.

Pankove, J. I. (1955). *Phys. Rev.*, **140**, A2059.

Pankove, J. I. and J. E. Berkeyheiser (1962). *Proc. IRE*, 50, 1976.

Panish, M. B., J. J. Queisser, L. Derick and S. Sumski (1966). *Solid State Electron.*, **9**, 311.

Pfann, W. G. (1952). *Trans. AIME*, **194**, 747.

Pfann, W. G., and K. M. Olsen (1953). *Phys. Rev.*, **89**, 322.

Pierron, E. D., D. L. Parker and J. B. McNeely (1966). *Acta Cryst.*, **21**, 290.

Piesbergen, U. (1963). *Z. Naturforsch.*, **18A**, 141.

Pilkuhn, M. H., and H. Rupprecht (1966). *J. Appl. Phys.*, **37**, 3621.

Pilkuhn, M. H., and H. Rupprecht (1967). *J. Appl. Phys.*, **38**, 5.

Piller, H. (1966). *Proc. Intern. Conf. Phys. Semiconductors*, Kyoto, p. 206.

Pizzarello, F. A. (1963). *J. Electrochem. Soc.*, **110**, 1059.

Quist, T. M., R. H. Rediker, R. J. Keyes, W. E. Krag, B. Lax, A. L. McWhorter and H. J. Zeigler (1962). *Appl. Phys. Letters*, **1**, 91.

Richards, J. L. (1957). *J. Sci. Instr.*, **34**, 289.

Richards, J. L. (1960). *J. Appl. Phys.*, **31**, 600.

Richards, J. L., and A. J. Crocker (1960). *J. Appl. Phys.*, **31**, 611.

Robinson, P. H. (1963). *RCA Rev.*, **24**, 574.

Rosi, F. D., D. Meyerhofer and R. V. Jensen (1960). *J. Appl. Phys.*, **31**, 1105.

Rupprecht, H. (1967). *Proc. Intern. Sym. on GaAs, Reading 1966*, Inst. Phys. and Phys. Soc. London, pp. 57–61.

Schell, H. A. (1957). *Z. Metallk.*, **48**, 158.

Seltzer, M. S. (1965). *J. Phys. Chem. Solids*, **26**, 243.

Shaw, D. W. (1966). *J. Electrochem. Soc.*, **113**, 904.

Shaw, D. W., R. W. Conrad, E. W. Mehal and O. W. Wilson (1967). *Proc. Intern. Sym. on GaAs, Reading 1966*, Inst. Phys. and Phys. Soc. London, pp. 10–15.

Skalski, S. (1962). *Compound Semiconductors, Vol. 1, Preparation of III–V Compounds*, Eds. R. K. Willardson and H. L. Goering, Reinhold Publishing Corporation, New York, pp. 385–389.

Sokolov, V. I., and F. S. Shishiyanu (1964). *Fiz. Tverd. Tela.*, **6**, 328.⟨*(Soviet Phys. Solid State*, **6**, 265, 1964).

Spitzer, W. G., and J. M. Whelan (1959). *Phys. Rev.*, **114**, 59.

Stearns, R. I., and J. B. McNeely (1966). *J. Appl. Phys.*, **37**, 933.

Steinemann, A., and U. Zimmerli (1963). *Solid State Electron.*, **6**, 597.

Steinemann, A., and U. Zimmerli (1967). *Crystal Growth, Proc. of an Intern. Conf. on Crystal Growth*, Boston, 1966, Ed. H. S. Peiser, Pergamon Press, New York, pp. 81–87.

Stern, F. (1966). *Phys. Rev.*, **148**, 186.

Stober, F. (1925). *Z. Krist.*, **61**, 299.

Strack, H. (1967). *Trans. Met. Soc. AIME*, **239**, 381.

Straumanis, M. E., and C. D. Kim (1965). *Acta Cryst.*, **19**, 256.

Sturge, M. D. (1962). *Phys. Rev.*, **127**, 768.

Sze, S. M., and J. C. Irvin (1968). *Solid State Electron.*, **11**, 599.

Tausch, F. W., Jr., and A. G. LaPierre (1965). *J. Electrochem. Soc.*, **112**, 706.

Thurmond, C. D. (1965). *J. Phys. Chem. Solids*, **26**, 785.

Tietjen, J. T., and J. A. Amick (1966). *J. Electrochem. Soc.*, **113**, 724.

Turner, W. J., and W. E. Reese (1964). *J. Appl. Phys.*, **35**, 350.

Ukhanov, Yu. I., and Ru. V. Mal'tsev (1963). *Fiz. Tverd. Tela.*, **5**, 2926. (*Soviet Phys. Solid-State*, **5**, 2144, 1964.)

Van de Pauw, L. J. (1958). *Philips Res. Rep.*, **13**, 1.

Van Muerch, W. (1965). *Solid State Electron.*, **9**, 619.

Vieland, L. J., and T. Seidel (1962). *J. Appl. Phys.*, **33**, 2414.

Wagner, R. S., and W. C. Ellis (1964). *Appl. Phys. Letters*, **4**, 89.

Weinstein, M., H. E. LaBelle, and A. I. Mlavsky (1966). *J. Appl. Phys.*, **37**, 2913.

Weisberg, L. R., F. D. Rosi and P. G. Herkart (1960). *Properties of Elemental and Compound Semiconductors*, Ed. H. C. Gatos, Interscience Publishers, New York, pp. 25–66.

Weisberg, L. R., and J. Blanc (1963). *Phys. Rev.*, **131**, 1548.

Welker, H. (1952). *Z. Naturforsch.*, **7a**, 744.

Welker, H. (1954). *Physica*, **20**, 893.

Whelan, J. M., J. D. Struthers and J. A. Ditzenberger (1960). *Properties of Elemental and Compound Semiconductors*, Ed. H. C. Gatos, Interscience Publishers, New York, pp. 141–154.

Whelan, J. M., and G. H. Wheatley (1958). *J. Phys. Chem. Solids*, **6**, 169.

Whitaker, J., and D. E. Bolger (1966). *Solid State Commun.*, **4**, 181.

White, J. G., and W. C. Roth (1959). *J. Appl. Phys.*, **30**, 946.

Willardson, R. K., and W. P. Allred (1967). *Proc. Intern. Sym. on GaAs, Reading 1966*, Inst. Phys. and Phys. Soc. London, pp. 35–40.

Williams, E. W., and D. M. Blacknall (1967). *Trans. Met. Soc. AIME*, **239**, 387.

Williams, F. V. (1967). *Proc. Int. Sym. on GaAs, Reading 1966*, Inst. Phys. and and Phys. Soc., London, pp. 27–30.

Williams, F. V., and R. A. Ruehrwein (1961). *J. Electrochem. Soc.*, **108**, 177c.

Winogradoff, N. N., and H. K. Kessler (1964). *Solid State Commun.*, **2**, 119.

Wolff, G. A. (1962). *Compound Semiconductors, Vol. 1. Preparation of III–V Compounds*, Eds. R. K. Willardson and H. L. Goering, Reinhold Publishing Corporation, New York, pp. 34–47.

Wolff, G. A., and B. N. Das (1966). *J. Electrochem. Soc.*, **113**, 299.

Wolff, G. A., P. H. Keck and J. D. Broder (1954). *Phys. Rev.*, **94**, 753.

Woodall, J. M. (1967). *Trans. Met. Soc.*, *AIME*, **239**, 378.

Woodall, J. M., and J. F. Woods (1966). *Solid State Commun.*, **4**, 33.

5

The Technology of Gallium Arsenide Lasers

C. D. DOBSON

Standard Telecommunication Laboratories Ltd., Harlow, Essex

Contents

5.1   Introduction

We have seen in previous chapters that a gallium arsenide laser consists essentially of a *p–n* junction of small area constructed in a parallelepiped of material (Fig. 3.1). Two opposite faces of the device are polished flat and parallel and perpendicular to the junction plane to form an optical resonant cavity.

Semiconductor lasers must operate at high current densities and in order to keep the requirements down to an acceptable level devices are made small with typical junction areas of 0.05×0.02 cm^2.

Electrical contacts must be applied to two of the device faces and, since the power dissipation in a laser is high, it is desirable that these contacts also be designed to remove heat from the device.

The performance of a gallium arsenide laser depends largely on its operating temperature. A typical pulsed laser has a threshold current density of 1000 A/cm^2 and a quantum efficiency of around 35% at 77°K. A similar device designed for operation at 30°C would have a threshold of 40,000 A/cm^2 and a quantum efficiency of 20%.

The process required to make a gallium arsenide laser can be summarized as follows (Fig. 5.1):

[1] Select suitable gallium arsenide material.

[2] Prepare material in suitable slice form.

[3] Form a p–n junction in the material.

[4] Apply suitable low resistance electrical contacts to the p–n wafer.

[5] Construct suitable optical cavities with appropriate reflectivity to give laser dice.

[6] Mount dice in a suitable heat-sink.

Modifications to this outline process are needed for certain specialized devices and for some applications particular features of the construction must be emphasized. In the following sections these processes will be considered in more detail.

5.2 Material selection

The preparation and selection of suitable laser material is still a major problem. Although much has been discovered recently about some of the necessary requirements for laser material a sufficient specification of carrier concentration, crystallinity and purity has not been found. It is common for 60% of all apparently satisfactory ingots to be unsuitable for constructing lasers, but the reason for this low yield continues to elude workers in the field of laser research.

Single crystal gallium arsenide is commonly used for laser manufacture though it is possible to work with twinned and even polycrystalline material if the laser mirrors are mechanically polished. Neither pulled nor boat-grown crystals show any particular advantages. To obtain low threshold currents and high efficiencies it is important to specify to close limits the carrier concentration of the material.

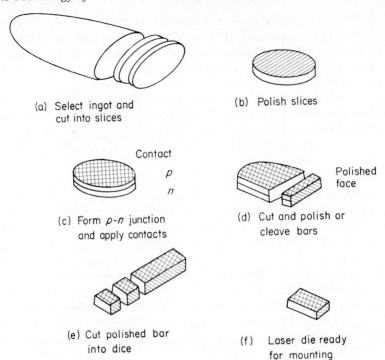

(a) Select ingot and cut into slices

(b) Polish slices

(c) Form *p-n* junction and apply contacts

(d) Cut and polish or cleave bars

(e) Cut polished bar into dice

(f) Laser die ready for mounting

FIG. 5.1. The processes required in the construction of a junction laser

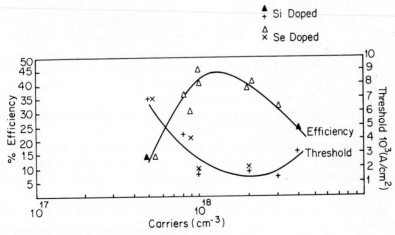

FIG. 5.2. The threshold and efficiency of diffused lasers as a function of doping level (77°K)

For lasers operating at 77°K a donor concentration of 2×10^{18} cm^{-3} appears to be optimum (Dobson, 1966) and silicon, selenium, tellurium and tin have all been used successfully. At higher temperatures higher donor concentrations are necessary to minimize the threshold current and there is some evidence that the choice of the donor element influences the delay which is observed between the application of the drive pulse and the emission of stimulated radiation. The use of silicon or tellurium appears to yield lasers with the shortest delays but with all dopants suitable heat treatment of the device is found to reduce the delays to values of 10 nsec or less.

Material with dislocation densities of 10^5 cm^{-2} and above is not suitable for making lasers. Twins in laser material often effect its cleaving properties and limit mirror formation to the mechanical polishing techniques. Polycrystalline areas in a laser crystal also spoil cleaved surfaces and can lead to uneven junctions whether formed by diffusion or by one of the epitaxial methods.

Laser material may also be produced by solution regrowth and vapour epitaxy and these techniques offer the advantage of better carrier concentration control and fewer defects. Mobilities in these materials are often higher than in the bulk-grown materials. However, mobility does not seem to be a significant factor in laser performance.

5.3   The preparation of laser material

5.3.1   Ingot cutting

Gallium arsenide ingots are mounted on thick glass plates with hard wax (Fig. 5.3) and cut into slices, usually about 0·5 mm thick. Ingots are most commonly cut with a diamond-impregnated peripheral saw. Saws as thin as 0·4 mm can be used when supported by metal sheets (Fig. 5.4a) and the thin blades give little wastage. The main advantage with peripheral saws is their ability to cut ingots at widely differing angles, often an important feature when cutting ingots which have crystal orientations well off the ingot axis.

The annular saw (Fig. 5.4b) has not the wide range of cutting angles of the peripheral saw due to the limited space inside the saw frame. It does, however, cut slices with the minimum of wastage since it uses a very thin (0·15 mm) annular saw blade which is tensioned by the saw frame.

Reciprocating multiblade saws use tensioned metal blades fed with an abrasive slurry (Fig. 5.4c). The method is relatively fast and produces very little damage in the slices. It is not suitable for cutting at large angles nor for cutting accurately to a chosen angle.

Cutting slices with an abrasive fed wire saw (Fig. 5.4d) also produces slices with very little damage. Wires only 0·1 mm in diameter can be used so wastage is slight. However, as with reciprocating blade saws, cutting on a particular orientation is difficult and operating the wire saw requires more skill and patience than either the peripheral, annular or reciprocating saws.

FIG. 5.3. A gallium arsenide ingot mounted for slicing

Unless gallium arsenide ingots are 'seeded' during growth the orientation of their crystal planes bears little relation to the ingot axis and slices cut from an ingot would have an arbitrary orientation. Since processes such as epitaxial growth and the use of chemical etchants require slices of specific orientation, gallium arsenide ingots must be orientated before they are cut into slices.

The spatial relationship of the {100}, {110} and {111} planes in gallium arsenide is shown in Fig. 5.5. Ingots are usually orientated so that when slices are cut their surfaces form (100) planes. This orientation is chosen because (100) surfaces are required for epitaxial growth processes. It will be shown in a later section that this orientation also makes it possible to construct laser cavities using the natural cleavage planes of gallium arsenide, i.e. the (110) planes, as the cavity mirrors.

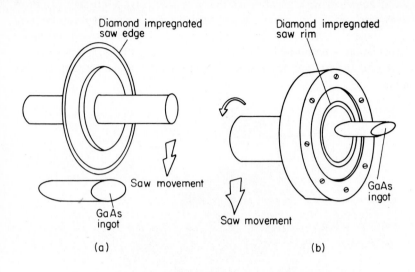

Diamond impregnated
saw edge

Diamond impregnated
saw rim

Saw movement

GaAs
ingot

GaAs
ingot

Saw movement

(a)   (b)

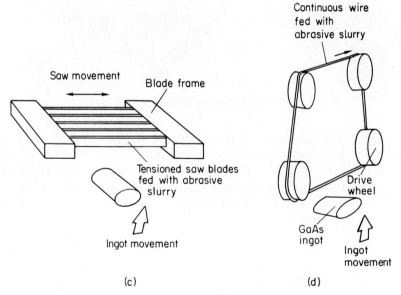

Saw movement

Blade frame

Continuous wire
fed with
abrasive slurry

Tensioned saw blades
fed with abrasive
slurry

Drive
wheel

GaAs
ingot

Ingot movement

Ingot
movement

(c)   (d)

FIG. 5.4. Sawing techniques used with gallium arsenide ingots: (a) peripheral
saw, (b) annular saw, (c) reciprocating abrasive saw, (d) wire saw

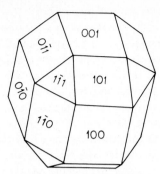

Fig. 5.5. The principal gallium arsenide crystal planes

Orientation is usually carried out by a combination of Laue photography and x-ray goniometry. A typical Laue photograph can be seen in Fig. 5.6.

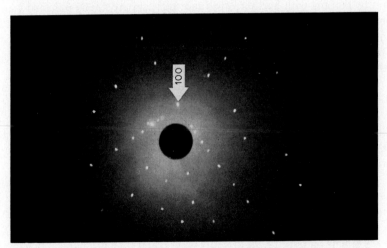

Fig. 5.6. Laue photograph from gallium arsenide ingot orientated near the (100) direction

The exposed spot produced by the (100) reflections has been marked and the fourfold symmetry which characterizes this crystal plane reflection can be seen. The technique is limited to an accuracy of about 1 degree by apparatus tolerances and film spot size. Greater accuracy ($\sim$ 1 minute of arc) can be obtained by using an x-ray goniometer.

5.3.2   The surface preparation of slices

After an ingot has been cut the slices are carefully removed from the glass cutting plate and washed in a detergent solution. Loose particles and grit can best be removed by ultrasonic cleaning although there is some risk of damaging the slices if this is carried out too vigorously. Any remaining wax can be removed by washing in a suitable solvent and, failing the usual organic solvents, hot glacial acetic acid can be used with great effect to remove most mounting waxes.

The sawing processes introduce damage into the surface of the slice. The extent of this damage varies with the technique used, but for practical purposes can be taken as extending a depth equal to four times the grit size used on the saw. Chips of material are often drawn through the saw cut and can cause additional damage which can be seen as score marks on the slice surface. The most damage-free cut surfaces are obtained using the abrasive saws described in the previous section.

Surfaces free of damage are required for diffusion and epitaxial processes. The damaged layer which was produced during the sawing stage is removed first by lapping with fine abrasives and the slices are finally polished using one of the three methods described below. Single slices can be lapped with an abrasive slurry on a glass plate. Using this technique it is difficult to obtain slices with an even thickness and machine lapping has considerable advantages.

Three polishing techniques are commonly used on gallium arsenide slices.

[1] Mechanical polishing can produce the flattest slice surfaces but a certain amount of damage is inevitably left in the surface. Diamond abrasives are less suitable than the softer alumina abrasives since they produce more damage.

One of the problems encountered in mechanical polishing is the tendency of the sharp edge of a slice to break giving chips which scratch the surface. This can be avoided if the sharp edges are removed by etching before polishing is commenced.

[2] Chemical polishing is in many ways easier than mechanical polishing but does not produce such flat surfaces. The slices must be carefully cleaned and degreased and again hot glacial acetic acid is a useful cleaning agent. A list of etches which can be used for chemical polishing is given in Table 4.2.

[3] A compromise between chemical and mechanical polishing has been developed by Sullivan and Kolb (1963). In this technique the lapped slices are spun on a rotating Teflon disc in a solution of bromine in methanol (0·05 volume %) giving an etch rate of approximately one micron per

minute. Reisman and Rohr (1964) have used a solution of NaOCl with a pellon cloth in the place of the Teflon disc. In this case polishing is carried out in two stages starting with a 10% solution and finishing with a more dilute 1% solution. The high concentration solution produces a smooth matt finish which rapidly develops into a polish with the 1% solution.

Surface profiles of gallium arsenide surfaces produced by sawing, lapping and polishing techniques are shown in Fig. 5.7.

5.4 The formation of *p–n* junctions for gallium arsenide lasers

5.4.1 *Diffused junctions*

One of the most critical stages in the production of a junction laser is the formation of the *p–n* junction. For successful laser action the junctions must be flat and have the correct impurity profile. These requirements have been met using both diffusion and epitaxial techniques.

In principle it would be possible to form suitable *p–n* junctions either by diffusing a donor into *p*-type GaAs or an acceptor into *n*-type material. Although the former technique has yielded successful lasers (Kelly, 1963) almost all GaAs lasers have been made by diffusing an acceptor into *n*-type material. Zinc is by far the most common acceptor diffusant although beryllium (Poltoratskii, 1965) has been used.

The diffusion of zinc into gallium arsenide has been extensively studied (Cunnell and Gooch, 1960; Weisberg and Blanc, 1963) and it is found that the diffusion profile does not follow the complementary error function impurity profile distribution which is obtained when the diffusion coefficient is a constant, independent of the diffusant concentration. The diffusion coefficient appears to be strongly concentration dependent, decreasing with reduced impurity concentration, and it has been possible to explain this behaviour on the basis of a model involving the equilibrium of substitutional and interstitial species of zinc. The electrical properties of the material are due to the substitutional atoms.

As a further complication, gallium arsenide will dissociate at the temperatures needed for diffusion so that it is necessary to perform the diffusion process in sealed ampoules. The dissociation of the material can be further reduced if the process is carried out in the presence of excess arsenic. This, however, affects the concentration of vacancies in the material and reduces the rate of diffusion.

Systematic studies of the diffusion process, such as those referred to, have acted as a guide in suggesting the most suitable diffusion schedules for laser junctions. However, in practice the diffusion schedules used in

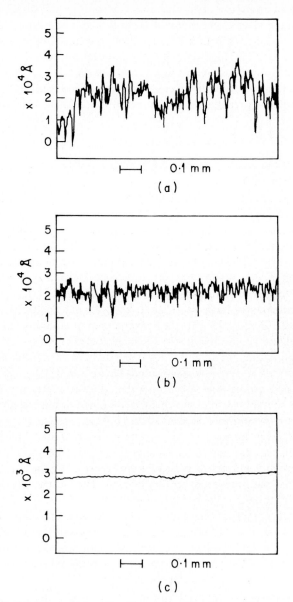

FɪɢG. 5.7. Surface profiles of gallium arsenide slice surfaces at various stages of preparation: (a) cut with diamond saw, (b) lapped, (c) mechanically polished (note that scale of (c) is 10 times that of (a) and (b))

different laboratories have usually been evolved on a somewhat empirical basis and there is no reason to suppose that the optimum diffusion conditions have yet been found (Marinace, 1963; Pilkuhn and Rupprecht, 1964).

The diffusion process is carried out by sealing the gallium arsenide slice, with weighed amounts of zinc and arsenic, in an evacuated silica ampoule. The ampoule is then heated in a furnace for the required time and temperature. A typical 2-h diffusion at 850°C, in a 20 ml ampoule containing 2 mg of zinc and 4 mg of arsenic, gives a junction 20 μ from the surface of the slice.

The silica tube used for diffusion requires careful cleaning and outgassing before diffusion as certain impurities cause considerable erosion of the gallium arsenide surface during diffusion. Care must also be taken to ensure a uniform temperature in the furnace otherwise thermal transport will cause the gallium arsenide slice to be further eroded.

Diffusion processes have been used with considerable success to make lasers for use at 77°K. At this temperature threshold current densities less than 10^3 A/cm^2 can be readily achieved and the resulting devices have high quantum efficiencies. Extensive development of diffused devices for use at room temperature, using multiple diffusion processes and annealing schedules, has also resulted in devices with threshold current densities $\sim 4 \times 10^4$ A/cm^2 and incremental quantum efficiencies of 20% at room temperature (Carlson 1967).

5.4.2   Epitaxial junctions

There is considerable evidence, both experimental and theoretical, to suggest that the achievement of low threshold current densities at room temperatures requires the highest possible doping levels in the *p–n* junction. Suitably high doping levels can be obtained using epitaxial processes in which material of one conduction type is deposited on a substrate of the opposite type and both vapour and liquid epitaxial techniques have been used (see Chapter 4).

Junctions formed by the deposition of zinc-doped layers in a halide transport system have been used to make lasers (Winogradof and Kessler, 1964) and such vapour-phase techniques are attractive since it is known that vapour-phase epitaxy can yield material with a minimum of unwanted impurities.

However, the development of solution regrowth techniques has probably had a greater impact on the epitaxial growth of lasers because the method is simple and lends itself well to producing heavily doped layers (Nelson, 1963).

Exact detail of the epitaxial process used varies in different laboratories, but the system is invariably a variant of that described by Nelson. A slice of *n*-type gallium arsenide is immersed in a 'melt' of gallium containing gallium arsenide and the dopant at 900°C. When zinc is used the concentration required is about 1 g of Zn per 25 g of gallium. The melt is allowed to cool slowly and as the temperature falls gallium arsenide grows epitaxially on the substrate.

In principle the epitaxial process could result in an abrupt junction. However, since the process occurs at an elevated temperature, diffusion of the dopants can occur giving a graded junction.

One of the possible disadvantages of epitaxial techniques is that the junction is formed at or near the original surface of the substrate, in a region containing work damage. This can to some extent be avoided if the melt is allowed to 'etch back' the substrate before growth commences. In this case the exact impurity profile in the junction region is difficult to specify since the material removed during the etch back can cause a compensated layer to be redeposited.

Once the junction has been formed excess *n*-type material is removed from the slice by lapping. It is not necessary to polish this lapped surface unless particularly thin slices (<0·1 mm) are required. In fact the lapped finish of one surface forms a useful method to distinguish the *n*- and *p*-type sides of the slice during subsequent processes.

TABLE 5.1. GaAs junction delineation etches.

Etch composition	Etching time at room temperature (sec)	Comments
HF 1 HNO_3 3 parts by H_2O 4 volume	5–10	Make in the order shown and use fresh
H_2O 10 HF 1 parts by H_2O_2 1 volume	15–30	Use with intense white light illumination. Deteriorates
HNO_3 100 parts by HF 1 volume	15–20	Somewhat erratic. Deteriorates
H_2O 500 CrO_3 250 parts by $AgNO_3$ 2 weight HF 1	2–4	Keeps well without HF. Add HF before use: remains usable for 48 h

Evaluation of a junction performance can only be done by completing test samples of lasers. However, it is useful to examine the depth of a junction and its flatness as a check on the junction-forming process. The scanning electron microscope is a valuable aid in this work as both the secondary electron and beam-induced current pictures give an indication of the position and flatness of a *p–n* junction. Satisfactory results can also be obtained by etching cleaved or bevelled sections. Cleaved sections have the advantage of simplicity and, when used with one of the etches shown in Table 5.1, give a reliable indication of junction depth and flatness. Bevelled sections are of value in examining shallow junctions and magnify junction irregularities.

5.5 Laser contacts

Laser contacts can be made either to the slice or to individual dice as part of the mounting procedure. Contacts made to the whole slice are easier to control and are more uniform so that this technique is to be preferred. *p*- and *n*-type materials do not have the same characteristics as regards electrical contacts and it is more difficult to make contact to *n*-type GaAs than to correspondingly doped *p*-type material.

TABLE 5.2. Materials used for evaporated contacts to galluim arsenide.[a]

Base metal	Eutectic temperature	Elements added	
		p-Type contact	*n*-Type contact
Gold	450°C	Zn Mn	Sn Ge Sn + In
Silver	650°C	Zn Zn + In	Sn Ge Ge + In Sn + In

[a] In some cases an alloy is evaporated, in others the components are evaporated as successive layers.

Electrolytic gold layers can be deposited on the GaAs slice using gold chloride solution with a trace of hydrofluoric acid. This makes an ohmic contact to *p*-type material if the slice is heated to about 500°C to alloy the gold into the slice surface. The contact to the *n* side of the slice, however, would be found to exhibit a high resistance which can be avoided if the gold is subjected to an electroless deposition of the donor tin before the alloying stage. Electroless nickel has also been used to make contacts but with unsatisfactory results even after heat treatment.

Probably the best contacts to gallium arsenide are made with vacuum-evaporated alloys. Table 5.2 shows some alloys which can be used to contact p- or n-type materials. The alloys are usually gold or silver based with a small amount (2–5%) of a donor or acceptor element to improve the electrical properties of the contact and sometimes a small amount (~ 10%) of indium to improve its wetting properties. The process is carried out by depositing about 0·3 μ of a suitable alloy onto the slice which is then heated in forming gas to a temperature at which an alloy with GaAs forms. Contacts made in this way require extremely clean surfaces and accurately controlled temperatures, otherwise the contacting alloy does not wet the GaAs surface evenly but forms hemispherical globules on the surface.

As an alternative to this process the alloy can be evaporated onto the slice while it is held at the alloying temperature. After an initial deposition of 1000 Å the slice is cooled and the contact thickness increased by a further layer of evaporated gold.

It should be noted that small area contacts, such as those formed by thermo-compression bonding, are not suitable for most lasers. Small area contacts have an appreciable spreading resistance and do not provide an effective means of removing heat from the device. Hence, for use in lasers operating at appreciable power levels, large area contacts with low electrical resistance must be used. With such contacts good thermal characteristics can also be obtained.

Contacts can also be made by alloying the laser to electroplated molybdenum, kovar or tungsten. In this case the laser is held in intimate contact with the gold or alloy plating on a metal stub and the whole assembly is heated in a reducing atmosphere until the gold alloys into the gallium arsenide, thus bonding it to the stub. This type of contact is easy to make but is difficult to control and non-uniform contacts, which have poor thermal characteristics, are often obtained.

5.6   Laser cavities

5.6.1   Mirror formation

GaAs has a refractive index of approximately 3·6. Hence light incident normally on a polished GaAs surface is partially reflected with a reflectivity of approximately 30%. However, since the gain per unit length in a gallium arsenide laser is high, this low reflectivity can be sufficient to make a device oscillate.

[i] *Polished cavities.* Laser cavities can be produced by polishing suitable surfaces perpendicular to the junction plane and parallel to each other. The slices are cut into bars with a width slightly in excess of the required cavity length. Three such bars are then clamped into slots in a polishing jig which is designed so that the slots are machined perpendicular to the jig surface (Fig. 5.8). The depth of these slots is slightly less than the required cavity length and the edges of the sawn bars are polished until they are level with the jig surface. The half-polished bars are then removed and the second cut faces polished in a similar jig with a slot depth equal to the required cavity length.

The flattest cavity faces are obtained using pitch laps (Dickinson, 1968) but even with the most careful polishing it is impossible to avoid some rounding at the mirror edges. This means that very shallow *p–n* junctions cannot be used with this technique.

The surface profile of a typical polished cavity is shown in Fig. 5.9.

[ii] *Cleaved cavities.* Gallium arsenide cleaves easily along the {110} planes and the resulting surfaces can be used to form the flat parallel faces of the laser cavity. The *p–n* junction must be orientated in a (100) plane so that cleaved surfaces are perpendicular to the junction.

Slices approximately 0·1 mm thick and less cleave very easily, but to cleave a chosen cavity length reliably requires considerable skill. The techniques for cleaving are numerous and each operator favours a particular method. Rolling a steel rod across the slice surface with its axis parallel to a cleavage plane gives a series of strips about as wide as the diameter of the rod. The pressure on the rod is important and the slice must be supported on a hard rubber pad. Cleaves are also to be made by making small scribe marks on the slice surface and then carefully bending the slice. Quite controlled cleaving is also obtained by pressing a diamond

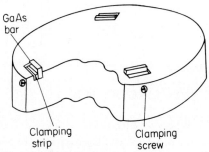

FIG. 5.8. Schematic diagram of a jig used for polishing laser cavities

stylus into the slice and using this method cleaved cavities only 50 μ long have been made on short strips.

A talistep trace of a mirror surface produced by cleaving can be seen in Fig. 5.9.

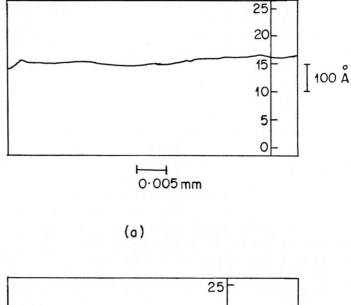

(a)

(b)

FIG. 5.9. Surface profile of (a) cleaved laser mirror, (b) polished laser mirror

5.6.2   Coating techniques

The basic laser previously described emits infrared radiation symmetrically from both Fabry–Perot mirrors. This is not convenient for most optical systems and in general undirectional lasers are more useful. To produce an undirectional device it is necessary to increase the reflectivity of one of the partially reflecting mirrors to a value approaching 100%.

The reflectivity of a GaAs surface may be increased simply by coating it with a metal film. The more commonly used metals with their reflectivities are shown in Fig. 5.10. Metal films of gold form good infrared reflectors and if evaporated onto the cold GaAs mirror they make a poor electrical contact which does not short circuit a significant portion of the diode forward current. Unfortunately these simple reflectors do not stand up to prolonged use and deteriorate after a period which depends upon the operating conditions. Operation at low power levels enables maximum life to be obtained from the reflector.

An insulating film interposed between the GaAs and the metal of the reflector produces a more reliable laser mirror. Evaporated layers of

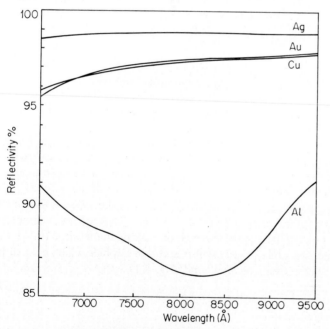

FIG. 5.10. The reflectivity of some commonly used metallic reflectors

silicon monoxide, silicon dioxide or aluminium oxide are suitable insulators. The thickness of the layer does not appear to be as critical as theory
might predict but is kept to about 300 Å in practice. The evaporated metal
is allowed to cover the junction but is masked across one edge of the die
to prevent short circuits developing between dice and the laser mountings
(Fig. 5.11a).

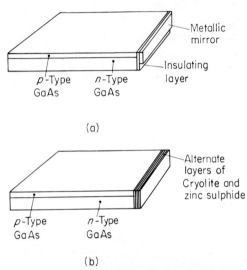

(a)

(b)

FIG. 5.11. The application of reflecting
coatings to laser surfaces: (a) insulated
metallic mirror, (b) dielectric mirror

Reflectors have also been made with aluminium paint (Kessler, 1967).
The paint is applied to the rear surface of the laser with a small brush. It is
not as effective as evaporated reflectors but is quick and simple to use. The
suspended aluminium particles which form the mirror are naturally coated
with hard aluminium oxide and therefore do not short circuit the junction.

Dielectric layers can also be used as mirrors. They are more complicated
to apply but are free from the risk of making short circuits. Alternate
quarter-wavelength layers of cryolite and zinc sulphide which have suitable
refractive indices can be used and Fig. 5.11b shows a diagram of the
dielectric arrangement.

5.6.3   Dice formation

The strips with their polished faces, which are the result of the processes
so far described, must be cut into dice of suitable width before they can

be mounted as individual devices. It is important that the side-edges of the dice should be poor reflectors otherwise there will be a tendency for the device to oscillate in undesired modes. The effective reflectivity of the cavity side can be reduced either by roughening the sides or by cutting them at an angle so that any reflected radiation is not directed back into the active region of the device. In most cases the roughness produced when the bar is sawn into dice is sufficient to suppress unwanted reflections and with some techniques the dicing process automatically leads to a cavity with sloping sides.

The bars can be cut into dice using a diamond saw or a wire saw. However, since the laser dice are very small, very high-quality saws are required otherwise the lasers will be cracked or chipped. Masking and etching techniques can be used to cut the bar into dice although this tends to be a laborious process.

5.7   Laser dice mounting

Laser dice are small, fragile and incapable of dissipating the heat which is generated during their operation. They must therefore be mounted to enable the heat to leave the dice more readily and to form a more robust and convenient assembly. The mean power operation of GaAs lasers is particularly dependent on the thermal performance of the mounting, and the thermal conductivity of some commonly used materials is shown in Figs. 5.12a and b.

Dice may be mounted by sandwiching the dice between two gold-clad metal discs and alloying the assembly together. The disc metal has to be a good expansion match to GaAs since the full thermal cycling range which the device must withstand ranges from the alloying temperature to the cryogenic operating temperature. Molybdenum, tungsten and kovar are suitable metals and the thermal expansions of a number of other mounting materials is given in Fig. 5.13. The gold on the clad discs is often doped with zinc for the p-side and tin for the n-side contacts. Often the alloying is controlled by visual inspection or with a thermocouple monitor at an alloying temperature of about 500°C in a hydrogen atmosphere, but it is difficult to produce good all-over bonding without including some void areas. Voids increase the thermal impedance of the device but it is possible to reduce them by using higher temperatures and slight pressure on the assembled parts. The layer of alloy formed between the mounting metal and the die is a poor thermal conductor and it is advisable to keep it as thin as possible. Fig. 5.14 shows a laser die mounted by alloying techniques.

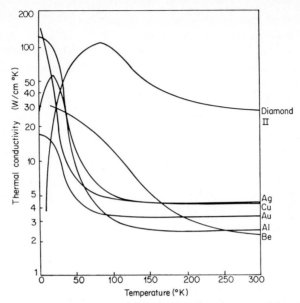

FIG. 5.12a. The thermal conductivity of some materials used
in laser construction

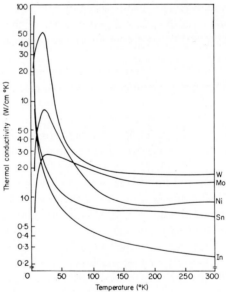

FIG. 5.12b. The thermal conductivity of some materials used
in laser construction

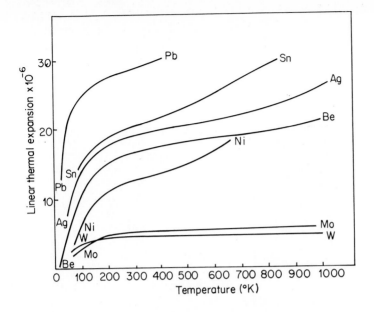

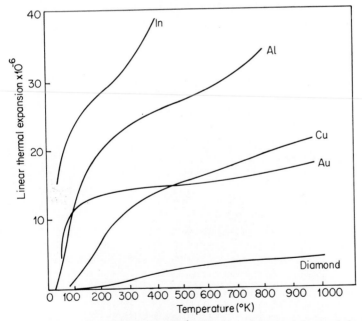

FIG. 5.13. The thermal expansion of some materials used in laser construction

Dice which have been contacted at the slice stage usually have very thin alloyed layers on their surfaces. Such dice can be soldered to metal mountings with lead–tin or pure tin solder. Pure tin is preferable since its thermal conductivity is higher. Fig. 5.15 shows a laser dice with one soldered contact.

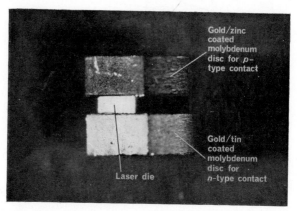

FIG. 5.14. A laser mounted by alloying between gold-plated contacts

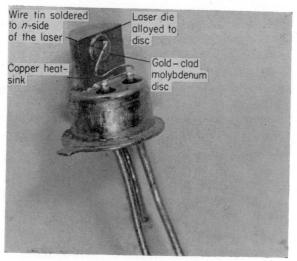

FIG. 5.15. A laser soldered to a transitor header

Thermally the most satisfactory method of mounting is to cold weld the die to a copper heat-sink. This is done by coating two copper blocks with indium and then pressing the die between them (Marinace, 1964). The bonding occurs at room temperature forming a good electrical and thermal contact between the die and its mounting. Large area contacts can be made this way without the formation of voids. The die used in the cold-welded mounting must have good ohmic contacts since indium does not make a satisfactory contact to GaAs unless heated to an alloying temperature of 400°C. Fig. 5.16 shows a typical assembly in which the two copper blocks are held together by a nylon screw.

FIG. 5.16. A laser die cold welded to copper blocks using indium

5.8 Special devices

5.8.1 Lasers for continuous operation

The normal operation of gallium arsenide lasers produces considerable power dissipation which raises the temperature of the device. As the temperature increases the threshold current rises and it is not generally possible to operate devices continuously.

It was shown in Chapter 3 that in order to achieve continuous operation attention must be paid to three features of the device. The construction of a C.W. laser thus relies on keeping the threshold current, electrical resistance and thermal impedance of the device as low as possible. The threshold current of a gallium arsenide laser decreases as its temperature is lowered and at the same time the thermal conductivity of most materials increases so that continuous operation is more easily obtained at lower temperatures. In fact the thermal conductivity of many materials shows a pronounced maximum near 20°K and laser threshold currents do not decrease rapidly below this temperature, so that in practice C.W. operation at liquid hydrogen temperatures can sometimes be achieved more readily than at liquid helium temperatures.

Engeler and Garfinkel (1964) have described a laser structure which is capable of 5 W continuous output when operating at 20°K. A schematic diagram of their device is shown in Fig. 5.17. The laser die is alloyed between two gold-clad tungsten discs which are separated by a slice of semi-insulating gallium arsenide. The use of ultra-pure tungsten ensures a high thermal conductivity for the heat-sinks and the use of semi-insulating gallium arsenide avoids any expansion mismatch in the structure.

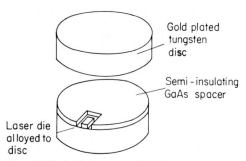

Gold plated tungsten disc

Semi-insulating GaAs spacer

Laser die alloyed to disc

Fig. 5.17. A high-power C.W. laser mount
(after Engeler and Garfinkel, 1964)

At 77°K laser threshold current densities and thermal conductivities become less favourable and continuous operation is only possible if careful attention is paid to all aspects of the laser design. Alloyed assemblies are not very satisfactory, but structures using cold-welded constructions have good thermal properties and can be used with advantage.

Above 90°K the techniques required for making C.W. lasers become quite sophisticated (Dyment and D'Asaro, 1967). Since heat can be removed most efficiently from narrow structures the width of the laser junction is kept as small as possible. Shallow junctions, only 2 μ deep, are produced by diffusion from dilute zinc solutions in gallium and the shallow junction reduces the thermal impedance between the junction and the heat-sink. The width of the junction is defined by the electrical contact to the *p* side which is in the form of a narrow gold stripe alloyed to the gallium arsenide (Fig. 5.18). The stripe, which is perpendicular to the (110) cleavage direction and is around 12 μ wide, is produced by etching through a silica layer deposited on the slice surface. The *p* side is then mounted by cold welding with indium to a heat-sink made from type II diamond coated with a metal film. Type II diamond is a high thermal conductivity material, and devices incorporating this diamond as a heat-sink have operated continuously at temperatures slightly in excess of 200°K. However, due to the narrow structures required, the devices can operate at only modest powers.

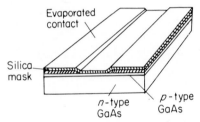

FIG. 5.18. A laser designed for continuous operation at 200°K (after Dyment and D'Asaro, 1967)

5.8.2 High power pulsed lasers

The peak power available from a cooled gallium arsenide laser is expected to be limited by the amplitude of the current pulse which is applied to the device and to the temperature rise during the pulse. In fact it is found that operation at high power levels can result in catastrophic damage to the device (Cooper, Gooch and Sherwell, 1966). Examination of the Fabry–Perot mirrors shows that damage occurs in the region of the junction (Fig. 5.19) and this is observed when output powers exceed 50 W/mm of junction width.

The laser fabrication can be modified to give higher peak power lasers by replacing the *p*-type material at the output face with low absorption *n*-type gallium arsenide. This is done by reducing the junction depth at the output face as shown in Fig. 5.20. Slices polished ready for diffusion are coated with 4000 Å of silica and a reference cleave is made across the slice.

FIG. 5.19. Damage on the surface of
a gallium arsenide laser

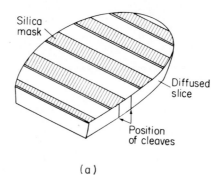

(a)

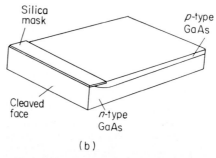

(b)

FIG. 5.20. A schematic diagram showing the form of a laser for operation
at high-power densities: (a) silica
diffusion mask, (b) section of laser die

Using a photographic mask and photoresist techniques a series of stripes is etched through the silica layer. The stripes are 2 mm wide with 0·25 mm spacing. The slice is diffused in the normal way and cleaved as shown in Fig. 5.20 with cleaves along the centre of each masked and unmasked region. The active region of the laser does not now extend to the output face of the cavity resulting in a lower power density and optical absorption at the cleaved surface. Output powers of 150 W/mm of junction width can be obtained using devices made in this way (Dobson and Keeble, 1967).

5.8.3 Room-temperature lasers

In principle any laser can be operated at room temperature if it is driven with a current pulse of sufficient amplitude. However, threshold current densities at room temperature are typically 25–50,000 A/cm^2 so that to make devices with reasonable threshold currents they must be small. A device with cavity dimensions 0·2 × 0·5 mm^2 would have a threshold current 50 A. Devices smaller than this are difficult to construct, but devices as small as 0·1 × 0·25 mm^2 have been made giving threshold currents as low as 10 A.

To obtain low threshold current densities at room temperatures high doping levels at the laser junction are required. These are often obtained using solution regrowth or liquid epitaxial techniques to produce the junction and in this way threshold current densities as low as 26000 A/cm^2 have been obtained (Nelson, 1963).

Lasers operating at temperatures above 200°K often show a delay between the application of the current and the onset of stimulated emission. These delays appear to depend to some extent on the donor element used with silicon and tellurium producing the smallest delays. The magnitude of the delay also depends on the way in which the laser is made with epitaxial techniques giving the smallest delays. However, the delays in diffused devices can be reduced by suitable annealing after diffusion.

5.8.4. Laser arrays

Single GaAs lasers are often used as small bright sources of radiation, but when brightness is less important an array of lasers can be used. The array can offer higher peak and mean powers than single devices in exchange for greater manufacturing complexity. The array is made up of individual dice usually connected in series to simplify the pulse generator requirements.

Simple laser arrays can be made by stacking laser dice together and soldering or cold welding them into a multielement die. A schematic diagram of the arrangement is shown in Fig. 5.21. The elements are electrically and thermally in series so that the mean power available is little more than that obtainable from a single device. The advantage of stacked lasers lies in the higher peak power available and in the relatively high brightness of this type of array.

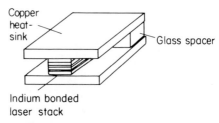

Fig. 5.21. A stacked laser array

A thermal improvement on the stacked device is shown in Fig. 5.22. In this array the laser dice are interleaved with metal sheets which are bonded to the die and to insulating spacers positioned on each side of the die. The heat is removed from the sheets by conduction through the large area insulators at the edges of the sheets. Indium welding suits this array construction particularly well and the arrangement allows a series–parallel removal of the heat produced in the devices while connecting the system electrically in series.

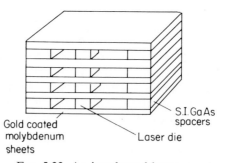

Fig. 5.22. An interleaved laser array

The removal of heat in parallel from an array can be achieved with the assembly shown in Fig. 5.23. Each laser die is mounted adjacently on an electrically insulating thermal conductor with dice bonded to metallized areas. The thermal conductor has metallized areas slightly larger than the dice distributed along the length. The metallizing may be sputtered or evaporated and the dice are connected in series by thermal compression bonded wires between die and metallizing. This design of array assembly has the lowest brightness of the three arrays described but has good thermal properties.

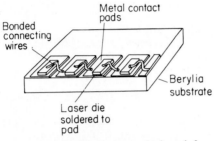

FIG. 5.23. A laser array designed for
high dissipation

5.9 Conclusions

Gallium arsenide lasers have been available for experimental purposes since 1962. The early devices operated at temperatures around 77°K, or lower, and only pulsed operation was possible. An important step towards the operation of devices at high duty cycles was the introduction of improved thermal structures using indium welding techniques (Marinace, 1964). Further development towards continuous operation at room temperature will probably depend on extensions of the techniques for making very narrow laser structures described by Dyment and D'Asaro (1967).

Continuous operation at temperatures approaching room temperature also requires devices with lower threshold current densities than are at present available. All the evidence suggests that this will require the high doping levels which can best be obtained using epitaxial techniques.

Research on gallium arsenide lasers has always been hampered by the poor reproducibility of the results obtained. This is still the situation and, although many of the conditions necessary for an efficient device can now be specified, the sufficient conditions have not yet been recognized. Until this can be done the development of gallium arsenide lasers will continue to be hampered.

References

Carlson, R. O. (1967). *J. Appl. Phys.*, **38**, 661.

Cooper, D. P., C. H. Gooch and R. J. Sherwell (1966). *J. Quantum Electron.*, **2**, 329.

Cunnell, F. A., and C. H. Gooch (1960). *Phys. Chem Solids*, **15**, 127.

Dickinson, C. S. (1968). *J. Sci. Inst.*, *(Ser.* 2), **1**, 365.

Dobson, C. D. (1966). *Brit. J. Appl. Phys.*, **17**, 187.

Dobson, C. D., and F. S. Keeble (1967). *Proc. Intern. Symp. on GaAs, Reading, 1966*. Inst. Phys. and Phys. Soc., London, p. 68.

Dyment, J., and L. A. D'Asaro (1967). *Appl. Phys. Letters*, **10**, 84.

Engeler, W., and M. Garfinkel (1964). *J. Appl. Phys.*, **35**, 1734.

Fenner, G. E. (1967). *Solid State Electron.*, **10**, 753.

Kelley, C. E. (1963). *Proc. I.E.E.E.*, **51**, 1239.

Kessler, H. K. (1967). *Proc. I.E.E.E.*, **55**, 99.

Marinace, J. C. (1963). *J. Electrochem. Soc.*, **110**, 1153.

Marinace, J. C. (1964). *IBM, J. Res. Develop.*, **8**, 543.

Nelson, H. (1963). *RCA Rev.*, **24**, 603.

Pilkuhn, M. H., and H. Rupprecht (1964). *Trans. Met. Soc. A.I.M.E.*, **230**, 296.

Poltoratskii, E. A. (1965). *Fiz. Tverd. Tela.*, **7**, 2231. (*Soviet Phys. Solid State*, **7**, 1798, 1966.)

Reisman, A., and R. Rohr (1964). *J. Electrochem. Soc.*, **111**, 1425.

Sullivan, M. V., and G. A. Kolb (1963). *J. Electrochem. Soc.*, **110**, 585.

Weisberg, L. R., and J. Blanc (1963). *Phys. Rev.*, **131**, 1548.

Winogradoff, N. N., and H. K. Kessler (1963). *Solid State Comm.*, **2**, 119.

6

Amplifiers, Switching Devices and Transient Effects

R. F. BROOM

Institute of Applied Physics, University of Berne

Contents

6.1　Introduction

What are the essential features of a GaAs laser which render it especially suitable as an amplifier and as a switching device? Since shortly after its discovery in 1962 researchers on three continents have studied the properties related to these modes of operation, it is therefore relevant to ask the reasons for such widespread interest. Two obvious advantages are the small size and high efficiency. A further asset is the relative ease with which it can be pumped and modulated. The gain in the active region is extremely high and may easily exceed 40 db/mm, allowing large power gains to be attained in structures of small dimensions. Because the radiative recombination takes place between extended energy bands rather than between discrete levels, the gain extends over a relatively wide spectrum, 10–20 Å, considerably easing the problem of matching one unit to another. Finally, and perhaps most important, is the very short stimulated recombination lifetime, making it possible to obtain a switching time of less than 1 nsec. This short lifetime and the broad gain profile allow, in principle, a bandwidth approaching 100 GHz.

Of course the GaAs laser in this context has also its limitations, some of which are common to GaAs lasers in general while others relate to a particular device. In the first category there remains the necessity for cooling to below 200°K to achieve continuous operation or, alternatively, the problem of high currents and low duty ratios at room temperature. In the second category two important considerations must be noted. First, the short spontaneous recombination lifetime results in a high rate of spontaneous emission and a correspondingly high background noise level in a GaAs amplifier. This sets the lower limit to the intensity of input signals which can be usefully amplified. For this reason therefore GaAs amplifiers are satisfactory only for relatively high level input signals. Second, the injected carrier density in the active region of the laser is quite high, greater than 10^{16} cm^{-3}, and in a switching operation this concentration must be changed significantly. To do so requires an energy input which is several orders of magnitude larger than for a conventional transistor logic element. Consequently power dissipation becomes a serious problem for a large number of units. It would therefore be wrong to suppose that GaAs laser switching elements can simply replace existing transistor logic.

In the following section the detailed behaviour of GaAs diode amplifiers will be discussed and the available results reviewed. Optical coupling between two GaAs lasers is treated in Section 6.3 and it will be shown how this can lead to several types of switching, either optically or electrically controlled. Experimental realizations of monostable and bistable devices will also be described. Finally, in Section 6.4, a brief review is given of some of the more recently observed high speed transient phenomena, such as spiking modes and resonance effects, giving a positive contribution to modulation at microwave frequencies.

6.2 Amplifiers

6.2.1 General considerations

The principle of a GaAs amplifier is essentially simple, to utilize the active region gain while at the same time inhibiting oscillation by elimination of the optical feedback. If α represents the sum of the losses experienced by a wave propagating along the active region due to absorption by free carriers, principally free holes, and by diffraction out of the active region then the following condition is obeyed at the laser threshold

$$R_1 R_2 \exp\left[(g-\alpha)\,2L\right] = 1 \tag{6.1}$$

in which R_1 and R_2 are the reflectivities of the diode Fabry–Perot mirrors, L the diode length and g the gain coefficient. For an amplifier the quantity on the left must be less than unity, or

$$L(g-\alpha) < \ln\left(1/\sqrt{R_1 R_2}\right). \tag{6.2}$$

Clearly therefore the reflectivity of the mirrors must be as small as possible in order that the single pass gain, G, given simply by

$$G = \exp\left[(g-\alpha)\,L\right] \tag{6.3}$$

can be large without the amplifier breaking into oscillation. Two methods exist for reducing the reflectivity to low values, less than 1%. That most generally employed is a dielectric coating a quarter wavelength in thickness and having a refractive index $n' = (n_1 n_2)^{\frac{1}{2}}$ where n_1 is the refractive index of GaAs and n_2 that of the external medium, usually unity. Silicon monoxide and calcium tungstate ($CaWO_4$) both satisfy this condition with a refractive index of 1·9. An alternative method is to lap or cleave the diode surface at an oblique angle to the junction plane to deflect internally reflected light out of the active region. In this connection Brewster's angle may be used to advantage provided the polarization of the incident signal is restricted to the appropriate plane.

Let us now consider the gain coefficient g in some detail. For a given diode it depends upon current density, wavelength, temperature and, as we shall see later, upon photon density. Assuming a uniform value across the thickness d of the active region Stern (1966a) has calculated the maximum gain (versus wavelength) as a function of nominal current density for a number of temperatures. His results for a GaAs diode having donor and acceptor concentrations of 1×10^{18} and 4×10^{18} cm^{-3}, respectively, are shown in Fig. 6.1. In the calculations it was assumed that the internal quantum efficiency η was unity, that all the radiation was confined to the active layer and that the thickness d was 1 μ. For other values of these parameters the current density is related to the nominal value by

$$J = J_{\text{nom}}\, d/\eta\Gamma \tag{6.4}$$

in which d is measured in microns and Γ represents the fraction of the propagating mode lying within the active region.

In order to obtain an idea of the magnitudes involved it is useful to consider now a simple example. Suppose a single pass gain of $G = 100$ is required from a diode of length $L = 3 \times 10^{-2}$ cm at a temperature of 80°K. Using (6.3) and assuming a value for α of 20 cm^{-1} one finds that $g = 173$ cm^{-1} and, from the curves of Fig. 6.1, this corresponds to a nominal current density of 2×10^3 A/cm^2. Finally, assuming $d = 1 \cdot 5\,\mu$ and $\Gamma = 0 \cdot 9$, we obtain $J = 3300$ A/cm^2, or about three times the threshold current for a laser of the same length.

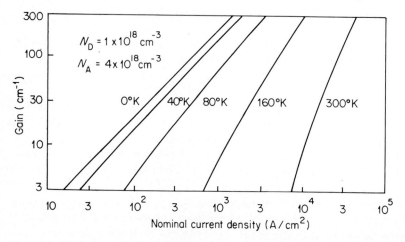

FIG. 6.1. Variation of gain coefficient g with nominal current density J_{nom} for a number of temperatures. J_{nom} is defined by relation (6.4). The donor and acceptor densities in the recombination region, assumed uniform, are taken to be 1×10^{18} cm^{-3} and 4×10^{18} cm^{-3}, respectively (after Stern, 1966a)

It will be noted from Fig. 6.1 that, at temperatures below 300°K, the relationship between gain coefficient g and current is approximately linear, in good agreement with experimental observation and with an earlier, more simple, analysis by Lasher (1963). Since g is proportional to current for small variations of the latter it follows, from (6.3), that the output from the amplifier $\mathscr{I}_0 G$ will have an exponential dependence on current. This is only true, however, at very low input intensities $\mathscr{I}_0$, the reason being that g is a sensitive function of photon density, decreasing as the latter is increased, due to depletion of the injected electron density. Consequently when the diode is long or the input intensity high g is no longer constant throughout the length but, because of the increasing intensity towards the output end, the gain coefficient is a decreasing function of distance and the expression for the single pass gain must be written as

$$G = \exp\left\{\int_0^L [g(x) - \alpha] \, dx\right\}. \tag{6.5}$$

Saturation of the gain by a high input makes the output become relatively independent of input and, furthermore, the output assumes a linear dependence on current. In other words, the majority of the pumping energy is diverted into the output. Under such conditions the amplifier has efficiency comparable to that of a laser driven at the same current density.

The effect of saturation may be seen quantitatively from the following relation derived by Stern (1966b) for the variation of light intensity with distance along the amplifier:

$$W + \ln W = 1 + (G' - 1)\alpha z. \tag{6.6}$$

Here the dimensionless light intensity $W = \mathscr{I}\tau_s/n_s$, where $\mathscr{I}$ is the photon flux in photons per square centimetre, τ_s the spontaneous recombination lifetime and n_s the injected carrier concentration at which the losses are balanced by the unsaturated gain. The quantity G' is related to the pumping rate by $G' = J\tau_s/edn_s$. This relation shows that when $W \ll 1$ the variation of intensity with length is logarithmic while when $W \gg 1$ the variation is clearly linear, both with length z and with current J, since W varies much faster than $\ln W$.

Saturation of the gain is well illustrated by the results of Crowe and Craig (1964) reproduced here in Fig. 6.2. A comparison of this curve with relation (6.5) by Stern (1966b) indicates that a reasonable fit is obtained for the following values: $\eta = 0 \cdot 7$, $\tau_s = 10^{-9}$ sec, $d = 1 \cdot 5\,\mu$ and $n_s = 1 \cdot 4 \times 10^{16}$ cm^{-3}.

An alternative description of the saturation of gain has been given by Nannichi (1966). He has shown that the gain coefficient at a given energy E is related to the overall recombination lifetime τ by the following:

$$g(E) = CJ\tau \tag{6.7}$$

in which

$$1/\tau = 1/\tau_{\text{spon}} + 1/\tau_{\text{stim}}$$

and C is given by

$$C = \frac{\sigma\pi c^3 h^3 \eta F(E)}{n^3 \, edE^2},$$

where σ is the absorption cross section of electrons and $F(E)$ their normalized distribution. The quantity C is dependent on the current density J through $F(E)$ but may be regarded as constant for a fixed current.

Using the dimensionless parameters $X = \tau_{\text{spon}}/\tau_{\text{stim}}$ and $Y = 1/(1 + X)$ it will be seen that $Y/\tau_{\text{spon}} = \tau$ and it follows that the variation of Y with X is equivalent to the variation of gain with photon density. Fig. 6.3 shows this variation and two distinct regions of operation, labelled I and III,

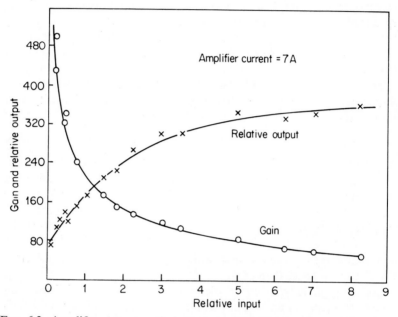

Fig. 6.2. Amplifier output and single pass gain G versus input intensity at low input levels. The amplifier current is 7 A and the temperature 77°K. Circles represent the measured gain and the crosses the relative values of the output (after Crowe and Craig, 1964)

which correspond to the domination of spontaneous or stimulated emission. The curve may be used to obtain values for τ_{stim} by comparison with the slope of the measured gain as a function of input intensity. Meaningful results can only be obtained, however, in regions I and II where the gain is relatively constant throughout the length of the diode.

6.2.2 Measurements on GaAs amplifiers

The arrangement used for the study of GaAs amplifiers generally has the form shown in Fig. 6.4. In order to avoid the necessity for high drive currents the diode is cooled to liquid nitrogen temperatures (77°K) by attaching it to a cooled heat-sink as shown, or to the inner container of a

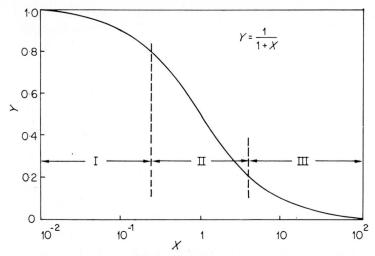

$$Y = \frac{1}{1 + X}$$

FIG. 6.3. Curve showing the variation of gain coefficient as a function of photon density, The dimensionless parameter $X = \tau_{\text{spon}}/\tau_{\text{stim}}$ and the overall recombination lifetime $\tau = Y/\tau_{\text{spon}}$. The regions labelled I and III correspond to the domination of spontaneous and stimulated emission, respectively (after Nannichi, 1966)

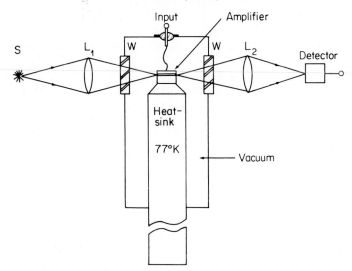

FIG. 6.4. Sketch of a typical arrangement for measuring the properties of a GaAs diode amplifier. Light from the source S, generally a GaAs laser, is focused on the anti-reflection coated surface of the amplifier by means of the lens L_1. Lens L_2 collects the output. The dewar windows W are also anti-reflection coated to prevent feedback into the amplifier

small cryostat. Optically flat anti-reflection coated windows allow the light to be coupled into and out of the diode. The problem of the input coupling is quite severe since the active region has a thickness of only about 1–2 μ. This requires all the components to be rigidly mounted and the elimination, as far as possible, of the effects of differential thermal expansion. The input light source is usually a GaAs laser. Unless it is placed within a few microns of the amplifier a lens is necessary to obtain effective optical coupling. Ideally the phase front from this lens should have the same form as that of the stimulated emission from the amplifier when operating as a laser. This implies a high quality lens of sufficient aperture to accept the beamwidth of the laser, about 25°. However, the insertion loss of the amplifier is generally greater than 10 db due to three principal reasons. First, the coupling loss caused by the lens; second, diffraction and absorption by free carriers in the active region of the amplifier ($e^{-\alpha L}$); and, third, inhomogeneities in the active region. The latter effect can be serious when the source is a laser showing filamentary behaviour, as the filaments may be directed onto regions of the amplifier having low gain.

Matching of the wavelength of the source to that of the amplifier is relatively simple due to the wide bandwidth of the latter, in the region of 20 Å. However, when a GaAs laser is used as input a spectral mismatch may arise when the amplifier is driven at a much higher current density than the laser, the reason being that the higher current in the former raises the quasi-Fermi level of electrons so that the amplifier reaches its maximum gain at higher energy than the laser. This effect will be more fully illustrated in a later section.

The properties of amplifiers have been studied by Coupland, Hambleton and Hilsum (1963), Crowe and Craig (1964) and by Crowe and Ahearn (1966). In all cases optical coupling was by means of a lens and anti-reflection coatings of either $CaWO_4$ or SiO were used to suppress oscillation. In the best units an increase in threshold current of up to ten times the uncoated value was obtained. Fig. 6.5 shows the single pass gain as a function of current for three different values of input intensity from the work of Crowe and Ahearn (1966). The curves labelled I and II were obtained with a very low input from the GaAs laser source, much less than 1 mW. They show the highest observed single pass gain of 2000. However, increasing the input intensity by a factor of three resulted in a drastic reduction in gain due to saturation (curve III). The peak gain in this latter curve is comparable to the results of Coupland, Hambleton and Hilsum, who, in addition, observed a linear relationship between current and gain. These results demonstrate the low input level at which saturation occurs and the linearity of gain versus current at higher inputs.

Coupling losses and the problem of mechanical stability are effectively eliminated by making the amplifier and laser driver initially from the same piece of material, then separating the two by cleaving after mounting on a semi-flexible support (Kosonocky and Cornely, 1968). By this means the air gap between laser and amplifier can be made as small as $0 \cdot 1 \mu$. It is not, of course, possible to coat the surface of the amplifier adjacent to the

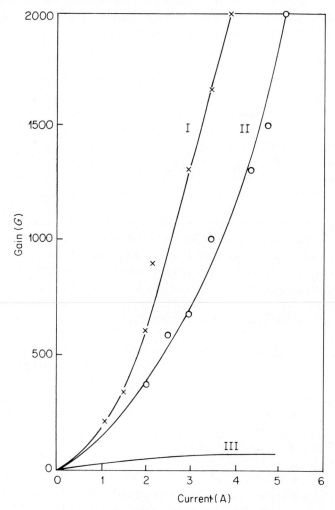

Fig. 6.5. Amplifier single pass gain versus drive current for three values of input intensity (after Crowe and Ahearn, 1966)

laser, but reasonable suppression of internal modes has been obtained by lapping the output end of the amplifier at an oblique angle of between 10 and 15° to the normal to the junction plane and perpendicular to the light direction. With such an arrangement it has been demonstrated that the quantum efficiency of the combination is close to that of a normal laser. A principal advantage of this system is the independent control of the spectrum and output power, the former being determined by the laser current and the latter by the amplifier current. Thus it is possible to obtain high output in a small number of oscillating modes.

6.2.3 Amplifier arrays

The principles mentioned at the end of the last section suggest how a further increase of power may be obtained by the use of a number of amplifiers all coupled in parallel to a single master oscillator. An array of GaAs amplifiers operating on this principle has been constructed by Vuilleumier and coworkers (1967). A diagram of their arrangement is shown in Fig. 6.6. Up to ten diode amplifiers were used, coupled to the single laser by a combination of spherical and cylindrical lenses, magnifying the x coordinate of the laser junction while preserving unity magnification in the z direction as shown. In this way the maximum amount of laser light was directed into the amplifiers. Anti-reflection coatings were applied to the latter in the usual manner, but an additional refinement was the inclusion of an isolator consisting of a $\lambda/4$ plate and a Glan–Thomson prism to deflect light from the amplifiers returned by the laser mirror out of the system. After amplification the outputs were collimated and passed through ten gas cells, one for each channel, in order to compensate for the path differences. After recombining the beams in a further optical system the gas pressures in the cells could be adjusted so that, in principle, all the beams were coincident in phase in addition to being coincident in frequency. Unfortunately, in practice, due to the small width of the emitting region of the laser ($\sim 30 \, \mu$) only three beams could be simultaneously phase locked in this manner. An overall gain of 13 db per amplifier was obtained at an operating temperature of 77°K. Measurement of the gain G at small values of input power showed close agreement with the theoretical logarithmic dependence of gain upon current.

6.2.4 Maximum obtainable gain

As the gain coefficient reaches a high value amplification of spontaneous emission (superradiance) becomes appreciable resulting in a reduction of the overall recombination lifetime. Reference to Fig. 6.3 shows that this in turn leads to a reduction in gain. Consequently as the diode current is increased the gain does not increase indefinitely but tends to a limiting value set by superradiance. Superradiance may be observed as a narrowing in the incoherent emission spectrum and also as a sharp increase in the noise output. An exact estimate of the limiting value of the gain is complicated by the geometry of the active region, but from the experimental results of Crowe and Ahearn (1966) an approximate value for the maximum gain coefficient g of 250 cm^{-1} may be deduced.

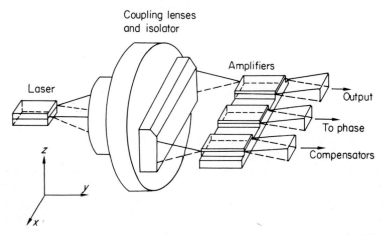

Fig. 6.6. Diagram showing an array of amplifiers coupled to a single GaAs laser. Only three amplifiers are shown for clarity. Details of the coupler and isolator are explained in the text (after Vuilleumier and coworkers, 1967)

6.3 Switching devices

6.3.1 General principles

The results described in the earlier part of this chapter have shown how the gain and its dependence on current is a rapidly varying function of the photon density. This fact may be exploited as a means for controlling the output of a laser through optical coupling with the radiation from a second laser. The latter radiation may then be used to lower the population inversion in the first laser, thus lowering the gain, and if the latter is not driven too far above threshold, reducing the gain to below the level required for oscillation. As a result the modes characteristic of the quenched laser are extinguished and only the amplified modes of the second laser, the emitter, appear in the output.

The cavity of the emitter may be aligned parallel or perpendicular to that of the laser to be quenched. In the latter arrangement there is the advantage, however, that the quenching light does not lie in the same direction as the switched output and hence may be easily eliminated from the detected signal. Both of these configurations have been demonstrated by Fowler (1963, 1964) using two separate lasers mounted on a common header. The power transfer ratio, defined as the ratio of the change in output power to the emitter power, was unfortunately considerably less than unity owing to the rather high coupling losses.

Much higher efficiencies may be obtained by incorporating both lasers in a single piece of GaAs. Since the light always remains within the material and as both active regions may be formed in the same junction, problems of alignment and diffraction loss are eliminated. Such structures have been studied by many authors, for example Kelly (1965), Kosonocky (1965), Kawaji (1964, 1966) and Yonezu (1967). A typical arrangement is shown in Fig. 6.7a. Two junctions of different lengths are diffused into a single substrate and are separated by a short region, about 100 μ in length, of the n-type substrate. Subsequently all four sides are cleaved, but two opposite sides are roughened, as shown, to prevent oscillation in a direction perpendicular to the length axis of the larger diode. When current is passed through the small junction (region I), laser action occurs only in a transverse direction due to the high absorption in the unpumped region of the longitudinal cavity. Increasing the current in region II, the quenching laser, depopulates the upper level in region I, reducing the number of photons in the transverse oscillation mode and simultaneously raising the gain and Q-factor in the longitudinal cavity. For equal current densities in both regions the Q of the latter cavity is the greater by virtue of its greater length and therefore in this condition the transverse mode is quenched and is replaced by oscillation in the longitudinal direction. The two situations are shown in Fig. 6.7b. The power transfer ratio has a value of about 0·7 at 77°K, an order of magnitude higher than for two separate lasers. Fig. 6.8 shows the output of the transverse laser, pulsed at a current of 3 A and the sharp fall in intensity when an equal current pulse was applied to region II.

A slightly different form of this device is obtained by roughening the entire area of the broad faces constraining laser action to the longitudinal direction only. For unequal current densities in the two regions, that carrying the lower current may assume the properties of a saturable absorber. Fig. 6.9 shows the construction of a typical laser of this type. Isolation between the two p-type regions, as with the bidirectional device, has been achieved in a number of ways, for example by oxide masking of a narrow strip across the substrate before diffusion, or by etching or sawing a groove in the p region to within a few microns of the junction leaving a

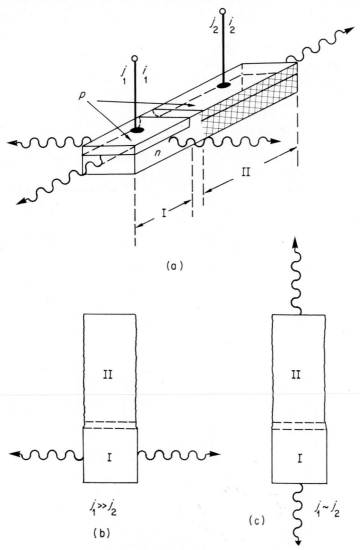

(a)

(b)

$j_1 \gg j_2$

(c)

$j_1 \sim j_2$

FIG. 6.7. Switching device employing two injecting regions and transverse and longitudinal cavities.

(a) Showing arrangement of two junctions and roughening of the sides of the longer junction.

(b) Transverse mode of oscillation when $j_1 \gg j_2$.

(c) Longitudinal mode of oscillation when $j_1 \sim j_2$ due to the higher Q-factor of the longer cavity

thin layer of low conductivity. The latter method is, in principle, the best since it preserves the dielectric waveguide properties of the junction in contrast to the greater diffraction resulting from two separate junctions.

Several authors have studied the steady state and switching behaviour of this device both experimentally (Kelly, 1965; Nathan and coworkers, 1965; Eliseev, Novikov and Fedorov, 1965; Basov and coworkers, 1965a, b)

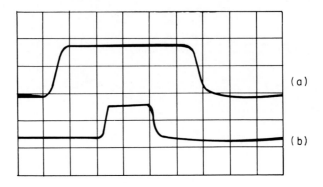

(a)

(b)

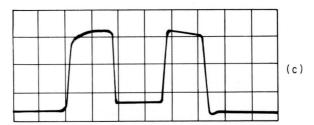

(c)

FIG. 6.8. Output waveforms obtained from the device of Fig. 6.7. Trace (a) represents the current through region I (1·6 A/division), (b) the current through region II (2 A/division). The transverse output is shown in (c) indicating the fall in output as oscillation is initiated in the longitudinal direction by the pulse (b). The time scale is 100 nsec per division and the temperature 77°K., from Kosonocky, W. F., Cornely R. H. and Marlowe, F. J., 'GaAs Laser Inverter', presented at the International Solid-state Circuits conference, February 17–18, 1965, Philadelphia. Reproduced by kind permission of the authors

and theoretically (Lasher, 1964; Zakharov and coworkers, 1966). The operation is quite different from that of the bidirectional device described earlier as here there is only a single cavity. Switching is observed in this case as a change in the wavelength of the oscillating modes as the ratio of the two currents i_1 and i_2 is altered. The mechanism responsible for this change in wavelength is described by the following analysis due to Zakharov and coworkers (1966).

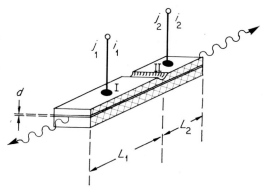

FIG. 6.9. Two-junction laser with unidirectional output. Under proper conditions of bias region I acts as an amplifier and region II as a saturable absorber

6.3.2 Simple theory of coupled lasers

The equations describing the time derivatives of the injected electron density and photon numbers $\dot{n}$ and $\dot{S}$, respectively, for two junctions of lengths L_1 and L_2 sharing the same cavity have the following form

$$\left.\begin{aligned}
\dot{n}_1 &= j_1/qd - n_1/\tau_s - S_1 g_1(E_1) - S_2 g_1(E_2), \\
\dot{n}_2 &= j_2/qd - n_2/\tau_s - S_1 g_2(E_1) - S_2 g_2(E_2), \\
\dot{S}_1 &= V_1 S_1 [g_1(E_1) + \gamma g_2(E_1)] - S_1/\tau_{\mathrm{ph}}, \\
\dot{S}_2 &= V_1 S_2'[g_1(E_2) + \gamma g_2(E_2)] - S_2/\tau_{\mathrm{ph}},
\end{aligned}\right\} \qquad (6.8)$$

where the subscripts 1 and 2 refer to the two regions I and II, τ_{ph} is the photon lifetime, d is the active region thickness, V its volume and $\gamma = L_2/L_1 \leqslant 1$. Solutions of these equations are most easily obtained for steady-state conditions although it will be seen later that transient solutions also exist which can lead to the generation of very short pulses. Meanwhile, however, we will simply put the left-hand sides of the equations (6.8) equal to zero and make some simplifying assumptions to enable a qualitative analytical solution to be obtained. First, assuming the diode to be at a temperature of $0°K$ implies that all electron states above the electron quasi-Fermi level F are empty while below F all states are filled. Second, assuming that all the injected electrons occupy a conduction band tail having an exponential density-of-states and that transitions occur between these states and a narrow acceptor level close to the valence band edge allows the electron density to be written as

$$n = \rho_0 E_0 \exp(F/E_0) \tag{6.9}$$

and the gain coefficient at a photon energy E to assume the simple form

$$g(E) = \frac{A\rho_0}{2} \exp(E/E_0), \quad \text{for} \quad E < F$$

and
$$\left.\begin{array}{c} \\ \\ \end{array}\right\} \tag{6.10}$$

$$g(E) = -\frac{A\rho_0}{2} \exp(E/E_0), \quad \text{for} \quad E > F.$$

Here ρ_0 and E_0 are constants describing the electron density-of-states in the conduction band and A is a constant of proportionality.[1]

Using Eqns. (6.8) and (6.10) the problem is now to find the dependence of threshold on the current densities j_1 and j_2 and on the length ratio γ and the similar dependence of the peak photon energies E_1 and E_2.

The simplest case is that in which j_2 is zero, i.e. region II is entirely absorbing and as S_2 also becomes zero the last of the equations (6.8) is eliminated. In this case, for operation close to threshold, the solutions for the photon number S_1 and energy E_1 are shown to be

$$S_1 = V_1 \left[\left(\frac{L_1 - L_2}{L_1} \right) \frac{j\tau_{ph}}{qd} - \frac{2E_0}{AV_1\tau_s} \right] \tag{6.11}$$

and

$$E_1 = F_1 = E_0 \ln \left[\frac{2}{A\rho_0 V_1 \tau_{ph}} \frac{L_1}{L_1 - L_2} \right]. \tag{6.12}$$

From (6.11) it is evident that as L_2 becomes an appreciable fraction of the total length of the diode j_1 must be increased to maintain S_1 at its threshold value. Also from (6.12) it will be seen that the photon energy is a minimum for a uniform current throughout the diode ($L_2 = 0$) and increases as L_2 increases. (It may be noted that these simple relations break down when L_2 becomes comparable to L_1.)

---

[1] The exact expressions for the gain coefficient are given in Chapter 2. They do not have the simple analytic form of (6.10) and the approximations used here cannot therefore be expected to yield quantitative values, they are nevertheless useful in obtaining a qualitative description of the behaviour.

When j_2 is finite S_2 is no longer zero and the complete set of the equations (6.8) has to be solved. However, at the low temperatures assumed here, the maxima of the gain coefficients g_1 and g_2 occur at energies coincident with the values of the electron quasi-Fermi levels F_1 and F_2 allowing the gain coefficients to be written in the form

$$\left.\begin{aligned} g(E_1) = g(F_1) = \frac{A\rho_0}{2}(1-\gamma)\exp(F_1/E_0), \\[2mm] g(E_2) = g(F_2) = \frac{A\rho_0}{2}(1+\gamma)\exp(F_2/E_0). \end{aligned}\right\} \tag{6.13}$$

Substitution in the equations (6.8) now yields the required solutions for the photon numbers S_1 and S_2:

$$S_1 = \frac{V_1}{2}\left[\frac{\tau_{\mathrm{ph}}}{qd}\left(\frac{L_1-L_2}{L_1}\right)(j_1-j_2) - \frac{4E_0}{AV_1\tau_{\mathrm{s}}}\left(\frac{L_1}{L_1+L_2}\right)\right], \tag{6.14}$$

$$S_2 = \frac{V_1}{2}\left[\frac{\tau_{\mathrm{ph}}}{qd}\left(\frac{L_1+L_2}{L_1}\right)(j_1+j_2) - \frac{4E_0}{AV_1\tau_{\mathrm{s}}}\left(\frac{L_1}{L_1-L_2}\right)\right]. \tag{6.15}$$

The first result (6.14) gives the variation of intensity at the photon energy E_1 when the current density j_2 is greater than the threshold value, obtained from (6.15) by equating the quantity in square brackets to zero, i.e.

$$j_2^{\mathrm{th}} = \frac{4E_0qd}{AV\tau_{\mathrm{ph}}\tau_{\mathrm{s}}(1-\gamma^2)} - j_1. \tag{6.16}$$

This shows that S_1 actually decreases as j_2 is increased and is finally quenched at the value of j_2 which reduces the quantity in square brackets in (6.14) to zero.

Finally, as the gain is a maximum at a photon energy $E = F$, we can obtain the difference in the two peak photon energies $E_1 - E_2$ directly from the expressions (6.13), namely

$$\Delta E = F_1 - F_2 = E_0 \ln\left(\frac{L_1+L_2}{L_1-L_2}\right). \tag{6.17}$$

An example of the spectral shift observed when the current density j_2 is varied is shown in Fig. 6.10, reproduced from the work of Zakharov and coworkers (1967). It demonstrates clearly the presence of two families of spectra corresponding to S_1 and S_2 and the transition to emission at lower energies when the current density becomes more uniform. Similar behaviour has been observed by Kelly (1965), Nathan and coworkers (1965), Kawaji (1964) and by Basov and coworkers (1965a). The magnitude of the shift is generally in the range of 1–5 meV, rather smaller than

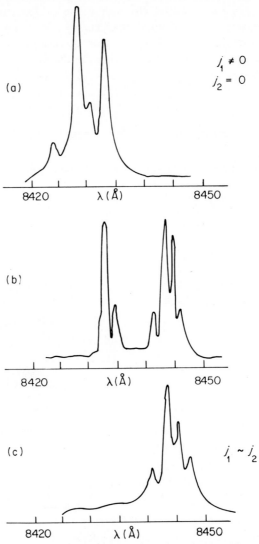

Fig. 6.10. Output spectra of the device of
Fig. 6.9 for various ratios of the current
density in the two regions, j_1/j_2. The wavelength reaches a minimum value when $j_2 = 0$
and a maximum when $j_1 = j_2$ (after Zakharov
and coworkers, 1966)

the value predicted by relation (6.17) assuming a value for E_0 of 20 meV (Stern, 1966a).

It will be seen from this analysis that, at threshold, the total current $i_1 + i_2$ is not constant as the current density in one region is varied, but increases with increasing asymmetry. This is verified by numerous measurements one of which is reproduced here from the results of Nathan and coworkers (1965) in Fig. 6.11.

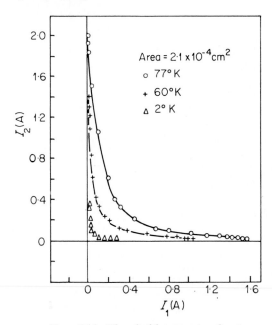

FIG. 6.11. Threshold currents of a two-junction device showing the increase in total current $i_1 + i_2$ with an increase in the asymmetry of the two current densities. Curves are for three temperatures of 2°, 60° and 77°K (after Nathan and co-workers, 1965)

6.3.3 Bistable devices

Bistable operation of the arrangement of Fig. 6.9 was first predicted and analysed by Lasher (1964) and, later, observed experimentally by Nathan and coworkers (1965). The principle depends mainly upon saturation of the acceptor to conduction band absorption in the unpumped region of the diode. If we suppose that the spontaneous emission from the pumped region excites sufficient electrons to the conduction band in the absorber to appreciably reduce the losses, then both the spontaneous emission from the pumped region and the emission due to recombination in the absorber has an opportunity of being reflected from the absorber mirror and returning to the active region of the pump. If this is sufficient to initiate oscillation the photon density increases in both regions so further reducing the losses in the absorber. Thus if the diode is driven initially at a current slightly

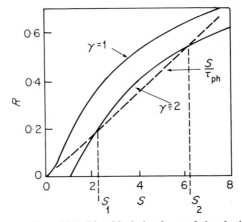

FIG. 6.12. Bistable behaviour of the device of Fig. 6.9. The ordinate is the total stimulated emission rate R, in a single mode, and the abscissa the photon density S. The full curves are the calculated values of R assuming low temperature, an exponential conduction band density-of-states and a constant valence band density-of-states, for two values of the length ratio γ. Cavity losses are represented by the straight line S/τ_{ph}. Bistability occurs when the latter cuts the curve for R at two points as explained in the text (after Lasher, 1964)

below threshold, it may be triggered into oscillation by either a small additional current pulse through emitter or absorber or by a light pulse into the absorber.

Lasher's analysis of the situation was based on similar rate equations to those of the preceding section. An example of his results is shown in Fig. 6.12. This gives the net rate of stimulated emission (R) as a function of the photon population of a single mode (S) for two values of the ratio γ, where $\gamma = L_{abs}/L_{em}$. The cavity losses may be represented by a straight line through the origin of slope S/τ_{ph}. Bistability is only possible if the latter straight line intersects the stimulated emission curve at two points. As an example consider the situation shown in the lower curve, for $\gamma = 2$. When S is less than S_1 the losses are greater than the emission rate and the output will decay to a small value. However, for S greater than S_1 the situation is reversed and the photon population will build up to the

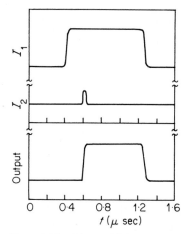

Fig. 6.13. Observation of bistable behaviour (schematic), showing switching to an 'on' state by the small current pulse i_2 through the absorber region (after Nathan and coworkers, 1965)

position of stable emission S_2. A minimum length of absorber is required for bistability, the exact value depending on temperature and the precise form of the conduction and valence or acceptor band density-of-states.

In the experimental observation of this effect by Nathan and coworkers (1965) the GaAs diode had dimensions of 225 μ in length by 40 μ in width with isolation between the two regions provided by etching. The resistance between the p regions was raised, by this means, to a maximum of 1000 Ω. At a temperature of 77°K and driving the emitter at a current (pulsed) of within 6% below threshold the device could be switched into a lasing state by a short current pulse through the absorber or, alternatively, by a light pulse from a second laser focused on the absorber junction at right angles to the cavity axis. Fig. 6.13 shows the two drive currents and the resultant output waveform with electrical switching. The switching time was less than 5 nsec and could be accomplished with a pulse as short as 1 nsec and with a current as low as one-tenth of the emitter current.

6.3.4   Further developments

The devices discussed so far have all been of a very simple type, having at most three terminals, they can respond only to a single input signal. Moreover, with the exception of the bistable device, the gain is generally less than unity and it is therefore impossible to couple optically several units together without the additional complication of amplifier stages. But it is essential for any opto-electronic computer application to have a number

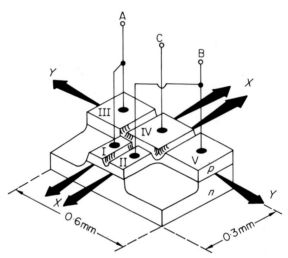

FIG. 6.14. An injection laser half adder. The 'T'-shaped recombination region is divided into five injecting contacts connected as shown and a constant bias current passed through contact C. Simultaneous pulses at A and B generate oscillation in the *Y* direction, but with non-coincidence the emission is in the direction *X* (after Kawaji, Yonezu and Nemoto, 1967)

of electrical inputs capable of switching the optical output between a number of alternative channels.

A significant development and good example illustrating a method of switching a light beam by a combination of inputs is the half adder of Kawaji, Yonezu and Nemoto (1967). The principle of operation is similar to that of the bidirectional device of Fig. 6.7a, described earlier, there being two orthogonal cavities of differing Q-factors. A diagram of the device is shown in Fig. 6.14. The rectangular GaAs substrate is cleaved on all four sides but the junction is etched to form a 'T' and five contact areas are isolated from one another by etching the four channels shown.

Region I is connected to region III making input A and region II is connected to region V for input B. In operation a constant bias current is passed through region IV optically pumping the remainder of the junction to an excitation close to threshold. When a current pulse is applied to A only, laser action occurs in the direction X, being inhibited in the other direction Y by the absorption of the region V. Similarly the direction of the output beam is unchanged for a pulse at B only. However, when current flows through both inputs the output is switched to the direction Y due to the greater length and consequently higher Q-factor of the cavity in this direction. Table 6.1 summarizes the operation.

TABLE 6.1. Summary of the output versus input characteristics of the GaAs laser half adder described in the text. (From Kawaji, Yonezu and Nemoto, 1967.)

Input		Output	
A	B	X	Y
0	0	0	0
1	0	1	0
0	1	1	0
1	1	0	1

The response time of the device was adequate to follow the 0·6 nsec rise and fall time of the drive pulses. Cooling to 18°K allowed the unit to be operated continuously.

6.3.5 Switching speed

Switching times of between a few nanoseconds and a fraction of a nanosecond have been mentioned, but generally these figures have been related to the rise and fall times of the drive current pulses or of the photodiode used to observe the output. For these reasons they evidently do not represent a lower limit to the switching times obtainable and so it is interesting to ask what this lower limit could be and what the factors are in its determination.

There are three basic time constants in the operation of a GaAs laser: first, the spontaneous recombination lifetime τ_s; second, the stimulated recombination lifetime τ_{stim}; and, third, the photon lifetime in the cavity τ_{ph}. The value of the first depends upon doping concentration and temperature; Dumke (1963) has calculated a minimum value of the order of 0·3 nsec while, in practice, a value of about 1 nsec is observed in typical diodes at a temperature of 77°K, increasing to 10 nsec at room temperature.

9

The stimulated lifetime, as discussed earlier, depends on the photon density within the cavity and hence on the degree of excitation; it may assume values greater or less than τ_s as indicated in Fig. 6.3, being smaller the higher the laser is driven above threshold. Since the overall recombination lifetime τ is given by

$$1/\tau = 1/\tau_s + 1/\tau_{stim} \qquad (6.18)$$

it follows that the last term dominates the response at high injection levels. This is illustrated by the results of Basov and coworkers (1966) for a diode at 77°K driven by a rectangular current pulse. The measured delay in the onset of stimulated emission was of the order of 1–2 nsec corresponding to the spontaneous lifetime. However, with the addition of a further sharp step to a drive pulse whose initial plateau was above threshold caused the stimulated emission to change to its new value in a very much shorter time, as small as 60 psec at a current of 17 times above threshold. This indicates that the response time of a GaAs laser is a maximum at currents below or close to threshold but at considerably higher currents can become very short indeed.

Similar considerations obviously apply to the absorber region of a GaAs switching device so that, in general, one may conclude that the response time will have a maximum value of the order of τ_s, but the higher the photon density and the higher the current the shorter the response time will be.

A lower limit to the response time is set by the photon lifetime within the laser cavity. This is defined by[2]

$$\tau_{ph} = \frac{nL}{c\varepsilon} \qquad (6.19)$$

in which n is the refractive index of the laser medium, L the cavity length, c the velocity of light and ε the fractional losses from the cavity during a single transit. For an uncoated GaAs laser of length 300 μ, ε has the value 0·6, neglecting absorption and diffraction loss, yielding a figure of 6 psec for the photon lifetime. It will be seen in the next section how the transient behaviour of a GaAs laser operated under certain conditions can approach this limiting time.

6.4 Transient effects: short pulses and high frequency modulation

6.4.1 Short pulse generation with the amplifier–absorber combination

GaAs lasers have recently been found to possess a number of properties leading to instabilities in their emission. Such mechanisms can lead to the

---

[2] See, for example, B. Lengyl (1966).

generation of extremely short pulses, can enhance the modulation depth obtainable at microwave frequencies and may even result in self-modulation of the output at frequencies in the gigahertz range. Some of these effects are similar to those observed in solid-state lasers; for example, spiking and Q-switching, but, because of the short time constants and small cavity dimensions of GaAs lasers, the pulse lengths may be very considerably shorter.

Let us consider first the amplifier absorber structure of Fig. 6.9. We have already seen how this device may be switched into an 'on' state by an additional small current pulse through the absorber region, allowing the photon density to build up to a sufficient value to saturate the absorber and sustain oscillation after removal of the triggering pulse. Now, however, we may note that the additional photon density partially saturates the gain in the amplifier section so that after an initial transition period the gain will fall slightly and, if not compensated by a decrease in loss in the absorber, the oscillation will again decay to a low value. Similar considerations may be expected to apply in the absence of a trigger pulse if the current is slightly *above* threshold. The cycle will repeat itself when the depopulated levels in the conduction band have been replenished by the injection current, resulting in the production of a train of regularly spaced short pulses. An analysis of the situation has been made by Rivlin (1967) and by Basov and coworkers (1967). The latter authors have used the rate equations (6.8) and the following instability condition

$$g_1 \frac{\partial g_1}{\partial n_1} + \gamma g_2 \frac{\partial g_2}{\partial n_2} < 0 \qquad (6.20)$$

in which g_1 and g_2 are the gain (or absorption) coefficients in the two regions, n_1 and n_2 the electron densities and γ the length ratio L_2/L_1. This relation implies that the combined gain coefficient

$$g = g_1 + \gamma g_2 \qquad (6.21)$$

must have a positive slope with respect to the photon density. But, since g is positive, the condition (6.20) can only be satisfied if either g_1 or g_2 is negative. In other words, there must be both an amplifying and an absorbing region. Applying the condition $dg/dS > 0$ to the steady-state solutions of the rate equations clearly gives an unstable solution as any small perturbation of the photon density is amplified rather than being damped.

The generation of short pulses by this mechanism has been observed by Drozhbin and coworkers (1967) using as detector a scanning image converter having a time resolution of 30 psec. The diode was operated with 10 msec pulses at a temperature of $77°K$. Their results for the measured pulse length and repetition frequency are shown in Fig. 6.15. It will be noticed that the pulse length decreases and the repetition frequency increases as the ratio i_1/i_1^{th} is increased. The current i_2 through the second junction was kept at a constant value of 8.0 A during these measurements.

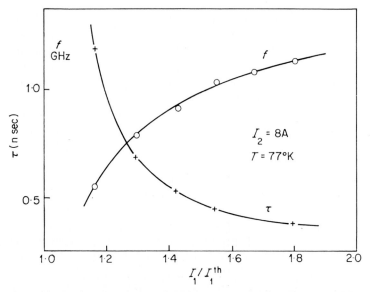

FIG. 6.15. Short pulse generation with an amplifier–absorber combination similar to the device of Fig. 6.9. The curves show the pulse length (crosses) and repetition frequency (circles) versus the ratio i_1/i_1^{th} for a constant current through the second junction (after Drozhbin and coworkers, 1967)

Close to the threshold for pulse operation it has also been found possible to trigger the pulses by means of a small additional current into one of the electrodes. Using this principle the pulse repetition rate has been synchronized to an external frequency of 500 MHz by Magalyas and coworkers (1967). In this case the output pulse length was only about 0.14 nsec. Fig. 6.16 shows the pulses, as observed with a scanning image converter, superimposed on a trace of the modulating waveform to illustrate the triggering of the pulse near the peak in the modulation current. Increasing the current caused triggering at an earlier point on the waveform while a still larger current caused bursts of pulses to be emitted at each positive half-cycle of the modulating current. Short pulses of similar form were actually first observed by Kurnasov and coworkers (1966) using a diode with a single contact. These authors were also able to generate single pulses of 0·2 nsec duration with a drive current pulse of 2·0 nsec in length (Kurnasov and coworkers, 1967).

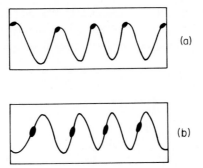

FIG. 6.16. Synchronization of short pulses at a repetition frequency of 500 MHz, (a) near threshold, (b) at higher current. The spots represent the pulses, observed with a swept image converter and the continuous line is a trace of the modulating waveform to show the constant phase relationship between the pulses and the current (after Magalyas and coworkers, 1967)

6.4.2 Spike modes

Short damped pulses are a familiar occurrence in the output of solid-state lasers. The effect is caused by a depletion of the inverted population at the antinodes of the standing electric field of a single longitudinal mode. The gain for this mode therefore falls below its initial value and conditions become favourable for oscillation in an adjacent longitudinal mode for which the antinodes are positioned to take advantage of the maximum gain over a large fraction of the total laser length. As a result the first mode decays and the second mode starts to build up, provided that diffusion of the excited states over a distance of half a wavelength is slow compared to the time required for a mode to build up in the cavity. A further consequence, at high pumping levels, is that a large number of modes are able to oscillate simultaneously.

This spiking behaviour was analysed for semiconductor lasers by Statz, Tang and Lavine in 1964 but had not at that time been observed experimentally, possibly because of the filamentary output from most diode lasers giving an averaging effect and the (predicted) short duration of the spikes. Recently, however, clearly resolved spikes at the start of laser action have been observed by Roldan (1967) in diodes made with the stripe geometry described by Dyment (1967). This structure is especially favourable for such measurements since it operates in a single transverse mode and

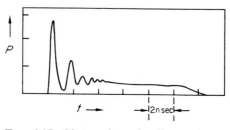

FIG. 6.17. Observation of spike modes at room temperature. The sharp rectangular drive current pulse is applied at a time near to $t = 0$ and terminates at the point where the output falls to zero (after Roldan, 1967)

without separate filaments. An example of the observed output is shown in Fig. 6.17 for a drive current pulse of 13% above threshold, 50 nsec in length and with the diode at room temperature. The initial pulse is indeed very short and is followed by a heavily damped train of pulses at decreasing time intervals. Solutions of the relevant rate equations, assuming a single longitudinal mode and negligible carrier diffusion, have been obtained by computer. The results, showing the variation of output as a function of time for three different values of drive current, are shown in Fig. 6.18. Comparison of the curves with those of Fig. 6.17 shows the excellent agreement with experiment despite the presence of several longitudinal modes in the observed output.

It is possible that this mechanism is also responsible for the 'pulse compresssion' effect mentioned at the end of the last section.

6.4.3 Q-Switching

Another totally different mechanism has been found to lead to a short pulse at the end of the current pulse (Ripper and Dyment, 1968). Nearly all the emitted energy is confined to a short spike, less than 0·4 nsec in width, during the trailing edge of the current pulse. The temperature and current at which the effect occurs are fairly critical, a typical value of the latter being 150°K. In addition, it is only observed in diodes showing an appreciable delay in the onset of stimulated emission, after application of the drive current, and a low transition temperature for the delay. An explanation of the effect has been given in terms of a double acceptor trap which may create one of two levels within the forbidden gap depending on whether it is singly or doubly ionized (Fenner, 1967; Dyment and Ripper, 1968). In its doubly ionized state the first level is near the valence band and

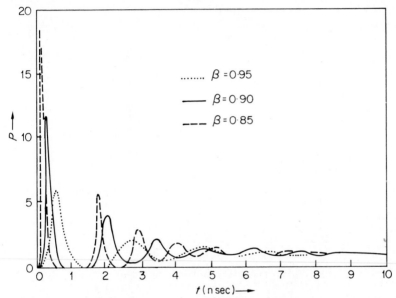

FIG. 6.18. Calculated spike output as a function of time for three values of the parameter β, related to the injection current, assuming a single longitudinal mode (after Roldan, 1967)

acts as an electron trap, to capture injected electrons. The capture of an electron by this centre, reducing it to a singly ionized state, has the special property of creating a second level close to or within the conduction band. This second level in turn behaves as an optical trap to capture electrons excited from the valence band. Its effectiveness depends on the proximity of the level to the electron quasi-Fermi level. Due to the presence of the optical trap laser action is suppressed at certain temperatures because of the additional optical absorption. However, when the injection current is reduced the lower trapping states empty, simultaneously removing the upper levels and effectively switching off the optical absorption. If this occurs in a time short compared to the spontaneous recombination lifetime there will be sufficient remaining inverted population to allow the diode to lase until the inversion is exhausted.

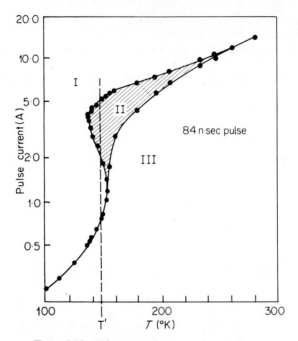

Fig. 6.19. Diagram showing the region within which Q-switching takes place, as a function of current and temperature. Normal laser action occurs in the region labelled I, spontaneous emission only is generated in region III and Q-switching, characterized by a short pulse of laser emission at the trailing edge of the current pulse, occurs within the shaded region II (after Ripper and Dyment, 1968)

The region over which this Q-switching takes place is represented by the experimental curves of Fig. 6.19. In region III only spontaneous emission is generated, in region I the diode lases normally, but within the shaded region II Q-switching takes place. It will be noticed that at some temperatures, such as that denoted by T', normal laser action can take place at currents both below and above the Q-switching region.

6.4.4 High frequency modulation

High frequency modulation of GaAs lasers, extending into the microwave region, is a relatively simple matter compared to the same exercise with other types of laser since the modulation may be applied directly via the injection current. High frequency modulation by this means was first demonstrated by Goldstein and Welch (1965) using waveguide techniques to attain frequencies in the X-band region. Still higher frequencies, 25 GHz, have been obtained by Ripper, Pratt and Whitney (1966) by the use of ultrasonic waves transmitted through the crystal. In the former case the modulation depth was quite small and the reason for this may be realized by the following argument for the case of a diode below threshold.

Suppose that an alternating current of the form

$$j = j_0(1 + A \cos \omega t)$$

is applied to the diode. At low frequencies the output will follow the input current variation both in amplitude and phase. However, when the frequency ω is raised until it becomes comparable to τ_s, the spontaneous recombination lifetime, the output will lag behind the input current and injected electron density with a phase angle δ given by $\tan \delta = \omega \tau_s$. Due to the finite value of τ_s the output is unable to keep up with the alternating current density. From this consideration one may easily obtain the ratio of the modulation depth at a frequency ω to the value at low frequency, it is

$$\frac{\mathcal{I}(\omega)}{\mathcal{I}(0)} = \frac{1}{(1 + \omega^2 \tau_s^2)^{\frac{1}{2}}}. \tag{6.22}$$

Thus for a diode at $77°K$, where $\tau_s = 10^{-9}$ sec, one may expect the modulation depth to begin to decrease at frequencies of greater than 100 MHz and about an order of magnitude less than this at room temperature. The situation for a laser is, however, rather more favourable than for a diode, especially if the former is driven at currents well above threshold so reducing the overall recombination lifetime considerably. Nevertheless, there still remains the problem of driving the low impedance diode with relatively heavy currents.

Two recently discovered effects can, however, contribute to an increase in the modulation depth at high frequencies; each results in actual enhancement of the modulation. The first effect arises from the photon lifetime within the cavity. Unlike the case of a simple diode where the output is always in phase with the electron density, this second time constant introduces a further phase lag between the photon density in the cavity and the injected electron density. Ikegami and Suematsu (1967) have analysed the problem with the rate equations (6.8) and shown that a resonance can occur, caused by these two separate time constants τ_s and τ_{ph}, which can lead to a considerable enhancement of modulation depth at certain frequencies. Their analysis leads to the following expression for the ratio $\mathscr{I}(\omega)/\mathscr{I}(0)$

$$\frac{\mathscr{I}(\omega)}{\mathscr{I}(0)} = \frac{\left(G + \frac{\tau_s}{\tau_{ph}}\right)\mathscr{I}_b}{\left\{\left[\left(G + \frac{\tau_s}{\tau_{ph}}\right)\mathscr{I}_b - \omega^2\tau_s^2\right]^2 + \omega^2\tau_s^2\left[1 + \left(G\frac{\tau_s}{\tau_{ph}} + 1\right)\mathscr{I}_b\right]^2\right\}^{\frac{1}{2}}}. \quad (6.23)$$

Here G is a dimensionless parameter related to the recombination rate and $\mathscr{I}_b$ is a function of the threshold current j_{th} and modulating current j_b, given by

$$\mathscr{I}_b = (j_b/j_{th} - 1). \quad (6.24)$$

When the ratio τ_s/τ_{ph} is greater than unity this expression leads to a maximum in the modulation depth versus frequency, as shown in Fig. 6.20. The position of the maximum depends mainly on τ_s and its magnitude on the ratio τ_s/τ_{ph}. Due to the approximations made in the calculation (similar to those of the analysis by Zakharov and coworkers (1966) in Section 6.3) this description of the effect must be regarded as more qualitative than quantitative, nevertheless the effect has been clearly observed by Ikegami and Suematsu (1968) in the frequency range 1–4 GHz. In addition the effect has been used to determine experimentally the photon lifetime, yielding a value close to 1 psec at a temperature of 77°K. As may be expected, this is rather shorter than the calculated value, using relation (6.19), presumably due to the effect of the diffraction and absorption losses which were ignored in the earlier estimate.

An analogous but more general analysis of the statistical fluctuations of the output of lasers has been carried out by McCumber (1966). His results for diode lasers indicate a fairly sharp peak in the spectral density of the output fluctuations at frequencies in the neighbourhood of 1 GHz. Experiments by D'Asaro, Cherlow and Paoli (1968) confirm these predictions with the observation of internally generated microwave modulation of the output in the frequency range 0·5–3 GHz in some diodes driven at two to three times threshold current. The modulated output was rather noisy and the frequency depended on both current and temperature. Stabilization at a given frequency could, however, be readily achieved by the injection of a small modulation current at the appropriate frequency.

A second mechanism contributing to the enhancement of modulation at high frequencies is the following: when the modulating frequency f

corresponds to the fundamental cavity frequency, $f_0 = c/2L$, for an empty cavity of length L or, as for a GaAs laser, including the refractive index n and dispersion $v(dn/dv)$,

$$f = \frac{c}{2L[n + v(dn/dv)]},$$ (6.25)

then the sidebands generated by the modulation, at frequencies $v_n \pm f$, where v_n is the optical frequency of the nth mode, coincides with the

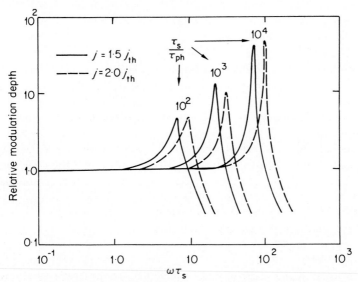

FIG. 6.20. High frequency modulation characteristics. Relative modulation depth $\mathscr{I}(\omega)/\mathscr{I}(0)$ as a function of the dimensionless modulation frequency $(\omega\tau_s)$ for two values of the average current j_0. The continuous curves are for $j_0 = 1.5j_{th}$ and the dashed curves for $j_0 = 2.0j_{th}$. The three sets of maxima correspond to three values of the ratio τ_s/τ_{ph} (after Ikegami and Suematsu, 1968)

frequencies of adjacent axial modes, v_{n-1} and v_{n+1}. The resultant interaction can create a definite phase relationship between the oscillating modes and this, in turn, leads to the development of a train of short pulses whose separation is $1/f$ and whose width is inversely proportional to the total number of oscillating modes. A good analysis is given by DiDomenico (1964). This is a well-established technique for generating ultra-short pulses with gas and solid-state lasers. For the case of a GaAs laser there are difficulties as ideally the gain or loss mechanism should be localized close to one of the mirrors and in addition the large dispersion makes it difficult to lock many of the modes. Nevertheless, it has been demonstrated that by 'tuning' the diode length according to relation (6.25) it is possible to obtain a considerable depth of modulation with relatively low powers at very high frequencies.

Takamiya, Kitasawa and Nishizawa (1968) have demonstrated such a mechanism at a frequency of 46 GHz. The modulated output frequency was too high to observe directly with a photodiode, but the sidebands generated could be easily resolved with a spectrometer. The results are reproduced in Fig. 6.21 which shows the spectrum with and without modulation. The diode length was 800μ corresponding to a mode separation of $1\cdot2$ Å, whereas the 46 GHz frequency is equivalent to $1\cdot1$ Å. It is evident that the modulation transfers a large fraction of the energy at the frequency of the central oscillating mode to the two adjacent modes. No such effect was observed with a short diode which did not approach the condition (6.25) indicating that the modulation depth was extremely small for this case.

A similar effect has been observed at relatively low frequencies, 300 MHz, with a diode coupled to an external cavity by means of a Brewster window (Broom and Mohn, 1968). Fig. 6.22 shows the variation of modulation depth at a constant drive current as the length of the external cavity was varied. Clearly defined peaks occur when the drive frequency is a multiple of the fundamental frequency of the cavity.

6.5 Summary

It is usual at this stage to make predictions about future developments but divinations of this sort are a hazardous exercise, probably best left in the hands of astrologers. It is, however, possible to give some comments on the more novel and significant aspects which have been discussed without becoming involved in crystal gazing.

As has been mentioned earlier, one of the problems associated with the use of amplifiers and switching devices is the optical coupling necessary between various units. This is especially acute when the light path cannot be constrained to a single p–n junction and when different coupled units have to be separated by an appreciable distance. Even although all the light can be transmitted from one active region to another there still exist losses due to differences in the internal electromagnetic mode patterns. A significant advance in this direction is therefore the ability to control the internal mode structure and to achieve reproducibility from one unit to another. This has been done recently by Dyment (1967) and Zachos (1968) with the development of a special diode geometry.

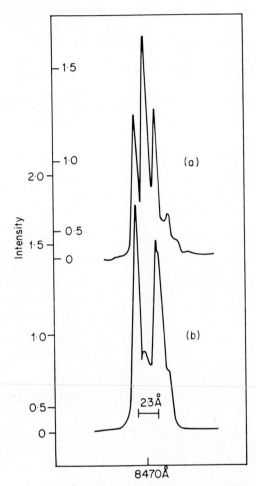

FIG. 6.21. Spectral change in the output of a GaAs laser modulated at a frequency of 46 GHz. (a) Spectrum without modulation, (b) spectrum with modulation, showing the transference of energy from the central mode to the two adjacent modes. The effect is only observed when the condition (6.25) is approxmated (after Takamiya, Kitasawa and Nishizawa, 1968)

It was stated in the introduction to this chapter that optical logic using GaAs lasers was unlikely to play any useful part in simply replacing existing transistor logic functions. Although the GaAs laser has fully demonstrated its ability to perform at high speed; by the generation of picosecond light pulses and amplitude modulation at microwave frequencies, the unique properties of such systems are more likely to lead to

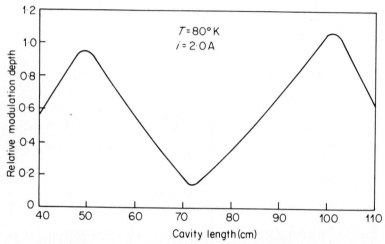

Fig. 6.22. Relative modulation depth of a GaAs laser operating within an external cavity, as a function of cavity length. The modulation frequency is 300 MHz (after Broom and Mohn, 1968)

totally new concepts in applications rather than extensions of established principles. Multichannel communication systems using pulse code modulation is one field of application where the use of such properties may be rewarding (Nelson, 1967). It is also likely that devices combining several of the effects we have described will be constructed in the form of a single unit.

Certainly it is clear that the study of transient behaviour in GaAs lasers is by no means exhausted. Future work, besides contributing to a fuller understanding of the presently known effects, will doubtless reveal new and interesting phenomena.

References

Basov, N. G., Yu. P. Zakharov, V. V. Nikitin and A. A. Sherenov (1965a). *Fiz. Tverd. Tela.*, **7**, 3128. (*Soviet Phys. Solid State*, **7**, 2532, 1966.)
Basov, N. G., Yu. P. Zakharov, V. V. Nikitin and A. A. Sherenov (1965b). *Fiz. Tverd. Tela*, **7**, 3460. (*Soviet Phys. Solid State*, **7**, 2796, 1966.)

Basov, N. G., Yu. A. Drozhbin, Yu. P. Zakharov, V. V. Nikitin, A. S. Semenov, B. M. Stepanov, A. M. Tolmachev and V. A. Yakovlev (1966). *Fiz. Tverd. Tela.*, **8**, 2816. (*Soviet Phys. Solid State*, **8**, 2254, 1967.)

Basov, N. G., V. N. Morosov, V. V. Nikitin and A. S. Semenov (1967). *Fiz. Tekh. Polup*, **1**, 1570. (*Soviet Phys. Semiconductors*, **1**, 1305, 1968.)

Broom, R. F., and E. Mohn (1968). *J. Appl. Phys.*, **39**, 4851.

Coupland, M. J., K. G. Hambleton and C. Hilsum (1963). *Phys. Letters*, **7**, 231.

Crowe, J. W., and R. M. Craig, Jr. (1964). *Appl. Phys. Letters*, **4**, 57.

Crowe, J. W., and W. E. Ahearn (1966). *J. Quantum Electron.*, **QE-2**, 283.

D'Asaro, L. A., J. M. Cherlow and T. L. Paoli (1968). *J. Quantum Electron.*, **QE-4**, 164.

DiDomenico, Jr., M. (1964). *J. Appl. Phys.*, **35**, 2870.

Drozhbin, Yu. A., Yu. P. Zakharov, V. V. Nikitin, A. S. Semenov and V. A. Yakovlev (1967). *Zh. E. T. F. Pis'ma.*, **5**, 180. (*J.E.T.P. Letters*, **5**, 143, 1967.)

Dumke, W. P. (1963). *Phys. Rev.*, **132**, 1998.

Dyment, J. C. (1967). *Appl. Phys. Letters*, **10**, 84.

Dyment, J. C., and J. E. Ripper (1968). *J. Quantum Electron.*, **QE-4**, 155.

Eliseev, P. G., A. A. Novikov and V. B. Fedorov (1965). *Zh. E.T.F. Pis'ma.*, **2**, 58. (*J.E.T.P. Letters*, **2**, 36, 1965.)

Fenner, G. (1967). *Solid State Electron.*, **10**, 753.

Fowler, A. B. (1963). *Appl. Phys. Letters*, **3**, 1.

Fowler, A. B. (1964). *J. Appl. Phys.*, **35**, 2275.

Goldstein, B. S., and J. D. Welch (1965). *Proc. I.E.E.E.*, **53**, 195.

Ikegami, T., and Y. Suematsu (1967). *Proc. I.E.E.E.*, **55**, 122.

Ikegami, T., and Y. Suematsu (1968). *J. Quantum Electron.*, **QE-4**, 148.

Kawaji, A. (1964). *Japan. J. Appl. Phys.*, **3**, 425.

Kawaji, A., H. Yonezu and Y. Yasuoka (1966). *Japan. J. Appl. Phys.*, **5**, 340.

Kawaji, A., H. Yonezu and T. Nemoto (1967). *Proc. I.E.E.E.*, **55**, 1766.

Kelley, C. E. (1965). *I.E.E.E. Trans. Electron. Devices*, **ED-12**, 1.

Kosonocky, W. F. (1965). In J. T. Tippett, D. A. Berkowitz, L. C. Clapp, C. J. Koester and A. Vanderburgh, Jr. (Eds.), *Optical and Electro-optical Information Processing*, M.I.T. Press, pp. 269–304.

Kosonocky, W. F., and R. H. Cornely (1968). *J. Quantum Electron.*, **QE-4**, 125.

Kurnasov, V. D., V. I. Magalyas, A. A. Pleshkov, V. G. Trukhan and V. V. Zvetkov (1966). *Zh. E.T.F. Pis'ma.*, **4**, 449. (*J.E.T.P. Letters*, **4**, 303,1966.)

Kurnasov, V. D., A. A. Pleshkov, G. S. Petrukhina, L. A. Rivlin, V. G. Trukhan and V. V. Tsvetkov (1967). *Zh. E.T.F. Pis'ma.*, **5**, 77. (*J.E.T.P. Letters*, **5**, 63, 1967.)

Lasher, G. J. (1963). *IBM J. Res. Develop.*, **7**, 58.

Lasher, G. J. (1964). *Solid State Electron.*, **7**, 707.

Lengyl, B. (1966). *Introduction to Laser Physics*, John Wiley and Sons, Inc., New York, Chap. 6., p. 237.

Magalyas, V. I., A. A. Pleshkov, L. A. Rivlin, A. S. Semenov and V. V. Tsvetkov (1967). *Zh. E.T.F. Pis'ma.*, **6**, 550. (*J.E.T.P., Letters*, **6**, 68, 1967.)

McCumber, D. E. (1966). *Phys. Rev.*, **141**, 306.

Nannichi, Y. (1966). *J. Appl. Phys.*, **37**, 3009.

Nathan, M. I., J. C. Marinace, R. F. Rutz, A. E. Michel and G. J. Lasher (1965). *J. Appl. Phys.*, **36**, 473.

Nelson, D. F. (1967). *J. Quantum Electron.*, **QE-3**, 667.
Ripper, J. E., G. W. Pratt, Jr., and C. G. Whitney (1966). *J. Quantum Electron.*, **QE-2**, 603.
Ripper, J. E., and J. C. Dyment (1968). *Appl. Phys. Letters*, **12**, 365.
Rivlin, L. A. (1967). *Zh. E.T.F. Pis'ma.*, **6**, 966. (*J.E.T.P. Letters*, **6**, 378, 1967.)
Roldan, R. (1967). *Appl. Phys. Letters*, **11**, 346.
Statz, H., C. L. Tang and J. M. Lavine (1964). *J. Appl. Phys.*, **35**, 2581.
Stern, F. (1966a). *Phys. Rev.*, **148**, 186.
Stern, F. (1966b). In P. L. Kelly, B. Lax and P. E. Tannenwald (Eds.), *Physics of Quantum Electronics*, McGraw-Hill Book Co., New York, pp. 442–449.
Takamiya, S., F. Kitisawa and J.-I. Nishizawa (1968). *Proc. I.E.E.E.*, **56**, 135.
Vuilleumier, R., N. E. Collins, J. M. Smith, J. C. S. Kim and H. Raillard (1967). *Proc. I.E.E.E.*, **55**, 1420.
Yonezu, H., A. Kawaji and Y. Yasuoka (1967). *Japan. J. Appl. Phys.*, **6**, 1018.
Zachos, T. H. (1968). *Appl. Phys. Letters*, **12**, 318.
Zakharov, Yu. P., V. V. Nikitin, A. S. Semenov, A. V. Uspenskii and V. A. Shcheglov (1966). *Fiz. Tverd. Tela.*, **8**, 2087. (*Soviet Phys. Solid State*, **8**, 1660, 1967.)

7

The Applications of Gallium Arsenide Lasers

K. G. HAMBLETON

Services Electronics Research Laboratory, Baldock, Hertfordshire

---

Contents

7.1 Introduction

The intention in this chapter is to give a practical guide to the application of gallium arsenide lasers. It is not an attempt to describe or catalogue the variety of uses already found for these comparatively new devices, but rather a general discussion of the principles involved in the design of laser systems, illustrated where possible by examples of practical applications. It is hoped that this will indicate the areas of choice and compromise open to the system designer, and while it will certainly not solve all his problems for him, at least it may help him to define them more clearly.

Most gallium arsenide laser systems can be divided into the same basic components. In sequence from the transmitter to the receiver they are the lasers, modulators or pulse generators, cooling if necessary, optics and detectors. These will be discussed in turn in the following sections, and finally the general features of laser communication and rangefinding systems will be considered.

7.2 Lasers

7.2.1 Review of laser properties

A brief study of the properties of gallium arsenide lasers described in Chapter 3 shows that they possess few of the attributes normally associated with lasers. The very small dimensions of the emitting aperture result in a diffraction limited beamwidth of several degrees which is very different from the narrow beams of gas or solid-state lasers. The spectrum of a gallium arsenide laser is not a single narrow line but consists of a complex family of modes separated by a few Angstrom units. Hence a GaAs laser is best regarded as merely a very bright small area source of infrared radiation with a spectral width of about 20 to 30 Å. Such a source can scarcely be called coherent since for a spectral width of 20 Å the coherence length is only a fraction of a millimetre. The interaction between the various modes from different portions of the emitting area can appreciably modify the output beamwidth, and a simple diffraction pattern is rarely obtained in practice.

Although gallium arsenide lasers are very imperfect from an optical point of view they have many important properties which render them attractive for practical applications. They can be modulated easily and at high speed; they are intrinsically efficient; they are extremely small and rugged; and they give long life and high reliability for relatively low cost compared with other types of laser. The lack of coherence is not a serious disadvantage as it is difficult to visualize many systems which could make full use of a coherent source. Coherence is very soon destroyed on transmission through the atmosphere by turbulence due to pressure and temperature variations, and even over relatively short distances the signals received are much less than expected (Rosner, 1968). Apart from possibly the important field of communications through space, coherent laser systems may well be restricted either to laboratory conditions or to applications where the transmission medium is artificially controlled. This implies some form of glass fibre with very low optical loss or maybe an evacuated tube. Even in these cases, maintaining the coherence over long distances will pose very serious mechanical and optical problems, quite apart from the difficulty of preserving the phase information throughout the detection process.

7.2.2 Brightness

The fundamental problem in any optical system is to direct sufficient power on to the detecting element, and a property of paramount importance as far as the source of radiation is concerned is brightness. If the source brightness can be increased by a factor of n^2 then the volume and weight of the associated optics can in general be reduced by a factor n^3 for the same performance, so that an order of magnitude increase in brightness, which would allow a $30 \times$ reduction in the space occupied by the optics, could well be the deciding factor as to whether a particular system were feasible or not.

Brightness is expressed in W/cm² steradian and depends on the optical power density at the emitting aperture and the beamwidth into which the power is emitted. The maximum optical power density at the emitting face of a gallium arsenide laser is in practice about 5×10^6 W/cm² since at higher power densities damage occurs at the polished surface of the laser cavity. If the emitting thickness of the $p–n$ junction is taken to be about 2μ this power limit corresponds to 100 W/mm of junction width. Peak power levels of this order have been reached in many laboratories with lasers operating at 77°K, and recently Nelson (1967) has approached it at room temperature, obtaining 70 W from a laser with a width of 0·08 cm.

At high peak powers most of the radiation from the laser is emitted into a beamwidth of $30° \times 30°$ or a quarter of a steradian so that for a power density of 5×10^6 W/cm² the brightness is about 2×10^7 W/cm² steradian. It must be emphasized that the figures in the above paragraphs are only approximate and represent typical rather than best results; nevertheless they are a good guide to present laser performances, and form a valid basis for the design of gallium arsenide laser systems.

It is instructive to compare the brightness of a laser with that of other sources of radiation, two of the more common being a tungsten filament and a carbon arc. If a tungsten filament is operated almost at its melting point the brightness is only about 100 W/cm² steradian and it would not last long under these conditions. The hottest carbon arcs reach about 1000 W/cm² steradian, but this is still at least 10^4 times less than the laser brightness. It must also be remembered that these other sources emit radiation over several thousand Angstrom units, whereas the laser has a spectral width of only 30 Å so that narrow band filters can be used to eliminate unwanted background radiation. Mercury or sodium lamps can be used with narrow filters, but the maximum brightness figures for these are only about 500 W/cm² steradian and 1 W/cm² steradian, respectively, so it is clear that the laser has no serious rival as a source for optical systems. Different types of lasers offer different properties, and ruby lasers are often used for long-range systems since they have a much higher peak power and brightness than gallium arsenide lasers, but they lack the modulation capability, compactness and simplicity of the latter.

7.2.3   Power output

Although the brightness of a gallium arsenide laser is limited to about 2×10^7 W/cm² steradian, the total power output will depend on the size of the device, and it is necessary to consider the factors that limit the dimensions of an individual laser.

The effect of cavity length on output power and efficiency was discussed in Section 3.3.4 where it was shown that at constant current density the output power increases with length up to a certain value but then remains constant, further increases in length merely resulting in lower efficiency. At 77°K the maximum useful length is about 2 mm for an unsilvered laser, or 1 mm if one face is totally reflecting. At room temperature the optimum length may be appreciably less than this. At first sight it appears that the emitting width of the laser could be increased indefinitely with a proportional increase in output power. However, apart from the obvious limits of gallium arsenide ingot size and inhomogeneity there is a fundamental limit due to the build-up of transverse modes. In practice, however, neither of these limits is reached and the laser width is restricted by the mundane problem of providing adequate pulses of drive current. As the size of the laser is increased larger currents are needed and the device impedance is proportionally reduced. Eventually a point is reached where it becomes impossible to drive sufficiently high-current pulses into the diminishing impedance with an adequate rise-time. Either a suitable current switch cannot be found or the effect of the stray circuit inductance limits the rise-time as the laser impedance decreases. Series resistance can be added to counteract the effect of the stray inductance, but this is obviously a very inefficient solution and the increased voltages required usually lead to equally difficult problems in other parts of the pulse generator. The problem is more serious for room-temperature operation since the current densities needed for efficient lasing are much higher at 300°K than at 77°K, and lasers designed to give maximum power at room temperature tend to be smaller than those intended for use with liquid-nitrogen cooling. Ideally a laser should be specifically designed and constructed for a particular system or class of system in the same way that different types of transistor are designed for different types of circuit operation.

7.2.4 Arrays of lasers

The small size of gallium arsenide lasers means that it is easy to increase the power input from a transmitter by using more than one device and building up a stack or array of lasing elements. It is clear from the previous paragraph that the individual lasers should be connected electrically in series, since parallel connection would be equivalent to using a single device of very large area and it would be virtually impossible to provide adequate current pulses. Obviously if a large number of lasers are needed an appropriate combination of series and parallel stacking can be used to match the resultant electrical impedance to the pulse generator.

The geometrical disposition of the laser elements will depend largely on the transmitter beamwidths required, and an array is especially useful when the output beam is to be wider than that from a single laser. A good illustration of this is a transmitter described by Birbeck (1968) which was built to take 20 lasers and give a beamwidth of 360° by 3° using collimation in one plane only. Fig. 7.1 shows a photograph of the transmitter with the cooling system and lens lifted clear so that the details of the laser mounting can be seen.

Although arrays of lasers are very convenient for generating large beamwidths they can also be used to give well-collimated beams. In Section 7.5.2, below, a transmitter is described which uses ten lasers at 77°K to give a peak power of 500 W into a beamwidth of only 5 by 2 milliradians.

These two transmitters were designed to use individually mounted devices, but an obvious extension of the technique is to fabricate all the lasers together in a single package. This gives a convenient compact source, and should eventually be cheaper since the encapsulation accounts for an appreciable fraction of the cost of a laser. It should also be possible to make and mount the lasers in multiple units instead of handling them singly. However, there are several disadvantages to this approach. The mean power of such an array will be severely limited by the problem of removing heat from the closely packed devices, particularly since series connection does not in general lend itself to good heat-sinking. Although the laser dice can be stacked very closely together, the space between the actual lasing junctions will mean a reduction in overall source brightness by a factor of at least a hundred. Finally, the system designer will be

Fɪɢ. 7.1. Wide beam transmitter using up to
20 gallium arsenide lasers

limited to the particular configurations produced by the device manufacturer, so that for special requirements the latter will have to develop special arrays and the advantage of low cost will disappear. Nevertheless, for low mean power work such arrays could be useful as cheap general-purpose sources. Units of up to 50 series-connected lasers have been produced in transistor-size packages, although the present price is dictated solely by the research and development costs, so it is inevitably very high.

7.3 Pulse generators

7.3.1 General requirements

One of the chief advantages which semiconductor lasers have over gas or solid-state lasers is that the light output can be modulated up to very high frequencies simply by varying the current through the device. At the present time C.W. operation is only possible with cooled lasers of relatively small junction area and low mean power output so most practical systems are pulsed.

Microwave modulation experiments discussed in Chapter 3 indicate that at temperatures of less than 100°K the light output should follow the current input down to pulse rise-times of 0·01 to 0·1 nsec. At higher temperatures, and particularly at room temperature, delays of up to a few hundred nanoseconds are sometimes observed between the current and light pulses (Fenner, 1967), but with suitable control of the type and level of the doping impurities these delays can be greatly reduced if not completely eliminated. Practical rise-times of less than 1 nsec are rarely encountered because of the difficulty of driving very fast high-current pulses into the very low impedance presented by the forward-biased laser diode. Hence for pulsed operation as well as for C.W. modulation the maximum speeds are limited by the design of the modulator and not by the intrinsic laser processes.

The simplest pulse generator is a d.c. supply and a switch and basically this is all that is required, although the switch must be turned on and off at suitable times to give the required pulse length and repetition rate. Because the pulse current is large and the laser impedance small the internal impedance of the supply must be low, and since laser systems usually operate at low duty cycles it is convenient to use a low-impedance storage element which is recharged at a relatively low current between successive pulses. A block diagram of a basic pulse generator is shown in Fig. 7.2 where a transformer is included between the switch and the laser to increase the output current and make the impedance matching easier. The repetition rate generator which is also used to synchronize the charging circuit with the switch is standard circuitry and need not be considered further. The charging network too is straightforward. Simple resistance or constant current charging can be used for low repetition rates although this is inefficient since power is continuously dissipated in the resistance. For higher duty cycles some form of synchronized pulsed charging is needed and, according to the speeds and currents required, this can be done with transistors, thyristors or gate-controlled switches (which have the advantage over thyristors in that they can be triggered to switch off as well as on). The other elements shown in Fig. 7.2 merit consideration in a little more detail so that the properties needed for use in gallium arsenide laser modulators can be emphasized.

7.3.2 Storage elements

The storage element must obviously be capable of holding sufficient energy for at least a single pulse, and the rate at which the energy is

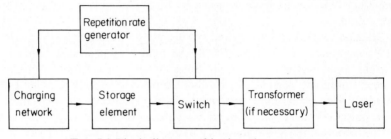

FIG. 7.2 Block diagram of basic pulse generator

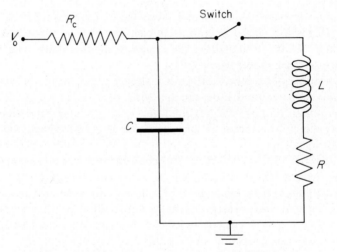

FIG. 7.3 Laser pulse circuit using condenser storage

released when the switch is closed (or in other words the pulse shape) is determined by the electrical structure of the element, which can be tailored to give a variety of pulse shapes as required. The most frequently used storage elements are either simple condensers for fast rising pulses or lengths of uniform transmission line for producing flat-topped pulses.

[i] *Simple condenser storage.* With a condenser as the storage element the equivalent circuit is a simple series *LCR* circuit as shown in Fig. 7.3

where the condenser has been charged to a voltage V_0 before the switch is closed at time $t = 0$. L and R are the total values of inductance and resistance in the circuit and include contributions from the laser, the leads, the switch and the condenser itself. Since the charging voltage V_0 is usually much larger than the laser forward voltage the latter can be neglected and the current can be determined by using the laser dynamic impedance in the values of L and R. Assuming that the switching action is instantaneous and that the condenser recharging current is small enough to be neglected, the expression for the laser current as a function of time is a standard formula with two distinct types of solution according to whether L is smaller or greater than $\frac{1}{4}CR^2$.

[1] $L < \frac{1}{4}CR^2$. This gives an over-damped pulse and the current I as a function of time t is

$$I = \frac{V_0}{kL}\exp\left(\frac{-Rt}{2L}\right)\sinh(kt), \qquad (7.1)$$

where

$$k^2 = \frac{R^2}{4L^2} - \frac{1}{LC},$$

therefore

$$\frac{dI}{dt} = \frac{-V_0 R}{2kL^2}\exp\left(\frac{-Rt}{2L}\right)\sinh(kt) + \frac{V_0}{L}\exp\left(\frac{-Rt}{2L}\right)\cosh(kt). \qquad (7.2)$$

Inserting $t = 0$ into Eqn. (7.2) shows that the initial rate at which the current rises is V_0/L which indicates the importance of keeping the stray inductance as low as possible. It is also seen from Eqn. (7.2) that the maximum value of the current occurs at a time t_m given by $\tanh(kt_m) = 2Lk/R$.

[2] $L > \frac{1}{4}CR^2$. This gives an oscillatory solution and the current is

$$I = \frac{V_0}{nL}\exp\left(\frac{-Rt}{2L}\right)\sin(nt) \qquad (7.3)$$

where

$$n^2 = \frac{1}{LC} - \frac{R^2}{4L^2}.$$

In this case, too, the initial rate of current rise is V_0/L, but dI/dt is zero when $\tan(nt_m) = 2Ln/R$ and this has a multi-valued solution for t, the current passing through successive maxima and minima as it decays. This is undesirable for laser work as the negative values of current may cause permanent damage by driving the laser into reverse conduction.

A measure of protection against this ringing and other negative transients can be obtained by shunting the laser diode with a fast switching diode connected with opposite polarity, but this can only be effective if the inductance of the switching diode and its leads is less than that of the laser.

[3] Critical damping.   The transition between the two types of solution occurs when $L = \frac{1}{4}CR^2$ and this represents critical damping. The current is

$$I = \frac{V_0 t}{L} \exp\left(\frac{-Rt}{2L}\right) \tag{7.4}$$

and has a single maximum value

$$I_\mathrm{m} = \frac{2V_0}{eR} \tag{7.5}$$

which occurs at time

$$t_m = \frac{2L}{R}. \tag{7.6}$$

Again the initial rate of current rise is V_0/L.

Fig. 7.4 illustrates the forms of the oscillatory and critically damped solutions together with a simple RC decay curve which implies $L = 0$ and thus represents the maximum possible amount of over-damping. Although in all cases the current initially rises at the rate V_0/L it is obvious from Fig. 7.4 that a better measure of the pulse rise-time would be the time taken to reach the maximum current. The implication of this is most easily seen in the critically damped case, where Eqns. (7.5) and (7.6) show that increasing R can effectively reduce the pulse rise-time but only at the expense of current amplitude, whereas decreasing L gives a faster pulse with no such penalty. Increasing V_0 as well as R will restore the amplitude to its original value, and this technique can be useful in low power circuits although the extra dissipation is a serious drawback at higher powers. Nevertheless, if the inductance cannot be reduced, it is common practice to add sufficient resistance to reach the critically damped condition so that a 'clean' pulse with no negative transients is obtained.

[ii] *Transmission line storage.*   To obtain rectangular pulses a uniform transmission line forms the storage element, and a lumped constant LC line is often used. For short pulses a suitable length of coaxial cable may be better and if necessary several coaxial cables can be connected in parallel to give a line of lower impedance. To avoid spurious reflections the line impedance should be made equal to the total resistance presented by the load and the switch, and if the matching is correct the voltage that

appears across this total resistance when the switch is closed is half the line voltage, so the current can easily be calculated.

The pulse rise-time is determined by the values of L and R external to the line as it was with the simple storage condenser, and it is a reasonable approximation to assume that it is given by $L/2R$, so that again it is necessary to minimize the circuit inductance if fast pulses are required.

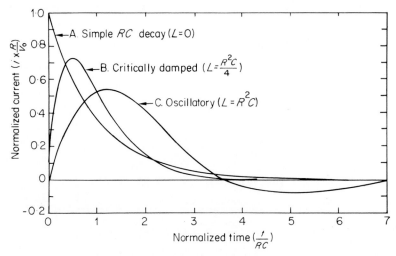

FIG. 7.4. Discharge current waveforms for series LCR circuit

The pulse shape can often be improved slightly by adjusting the values of the components at the beginning and end of the storage line to compensate for some of the parasitic circuit elements, and obviously a wide variety of pulse shapes can be obtained by using different component values along the length of the line.

7.3.3 Switches

For gallium arsenide laser applications the two most important properties of a switch are the closing speed and the current handling capability, although the resistance in the 'ON' state, the recovery time and the operating life must also be considered. A wide variety of switching elements has been used with lasers, and the best choice for a given application is determined by the requirements of rise-time, current, repetition frequency and the feasibility of using a pulse transformer between the switch and the laser.

[i] *Thyristors (or silicon-controlled rectifiers).* Thyristors are very useful
for general-purpose pulse generators as they can handle high currents and
give reasonably fast rise-times. The rate of change of current with time is
approximately 1000 A/μsec for a wide range of thyristor sizes, so that
100 A can be switched in 100 nsec. This is usually adequate for lasers at
77°K, although rather slow for room-temperature operation. The fastest
thyristors at present will switch in 30 nsec, but the current is limited to
about 40 A and it may be necessary to use a small-ratio pulse transformer

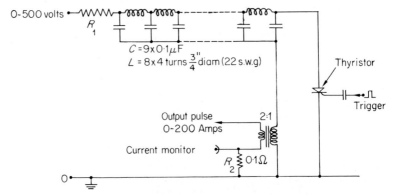

Fig. 7.5. Simple thyristor pulse generator circuit

to obtain sufficient pulse amplitude. Pulse repetition rates are limited by the
recovery time of the thyristor to less than 10^5 pulses per second, and the
recharging current must not exceed the thyristor holding current during
the recovery time.

Fig. 7.5 illustrates a simple thyristor circuit. The storage element is a
lumped constant *LC* delay line giving a pulse length of 5 μsec. Output
pulse currents from 0 to 200 A can be obtained by simply varying the d.c.
charging voltage, and rise-times of about 100 nsec can be achieved with
careful circuit layout and construction. The circuit was designed for
repetition rates below 100 pulses per second so that simple resistance
charging was sufficient. The charging resistance R_1 had a value of 10 to
20 kΩ depending on the holding current of the particular thyristor used.
The voltage across the monitoring resistance R_2 is given by:

$$V = IR_2 + L\frac{dI}{dt}, \tag{7.7}$$

where I is the current and L the inductance in the resistance itself. If accurate monitoring of very short pulses is required this inductance must be kept extremely small or the true current pulse will be masked by the inductive spike. In the present case the voltage across R_2 at full current output is 20 V, and the amplitude of the inductive spike is equal to this for an inductance of only 1 nH. This illustrates the magnitude of the problem, although with a pulse as long as 5 μsec the true value of the current is easily inferred from the flat top of the monitor pulse after the initial inductive spike.

Several thrysistors can be used in the output stage of a pulse generator, either in parallel to increase the current or triggered in sequence to give higher repetition rates (Dousmanis, 1966). With parallel operation it is usual to provide a separate storage element for each thyristor to ensure that the current is shared equally between them. The small internal resistance in the conducting state (which may be as low as 0·05 Ω) allows efficient operation at low voltages, and the ruggedness and reliability of thyristors make them ideal for use in laser systems, except where very fast pulse rise-times are required.

[ii] *Spark gaps.* Pressurized spark gaps were developed for use in radar modulators many years ago (Wilkinson, 1946). They can be used to produce pulses of up to 10^4 A with rise-times of less than 5 nsec (Adlam, 1960). Pulse jitter is bad even with triggered or three-terminal spark gaps, corrosion of the electrodes is difficult to prevent and high voltage supplies and insulation must be provided so the transmitters are considerably larger and heavier than those using solid-state devices.

[iii] *Mechanical switches.* Simple mechanical switches can provide fast high-current pulses but the repetition rates are limited to about 100 pulses per second. They are very efficient since their internal resistance is practically zero. Broom (1965) obtained currents of up to 2500 A with a rise-time of about 10 nsec using a coaxial configuration to minimize the circuit inductance. The contacts operated in air but arcing and corrosion were not serious problems since the pulse length was only 30 nsec and the storage element was completely discharged before contact-bounce occurred. Such pulse generators have operated for at least 10^8 pulses with only minor adjustments.

Mercury-wetted relays have also been used for driving lasers, but since the contacts are enclosed in a glass envelope, there is less scope for reducing the inductance. They can give rise-times of less than 1 nsec if a large series resistance is added but this is not practical at the high currents necessary for lasers. A typical relay pulse generator will give currents of a few hundred amps with a rise-time of 20 nsec at a repetition rate of a few hundred pulses per second.

[iv] *Discharge tubes.* Thyratrons and krytrons have both been used in laser pulse generators. They will operate at applied voltages of several kilovolts to give peak currents of up to 4000 A (Sullivan, 1964). If rise-times of less than 10 nsec are required appreciable series resistance must be added and the peak current will be reduced. Krytrons have the advantage over thyratrons of not needing cathode power supplies, but tend to suffer more from pulse jitter. Deionization times are a few hundred microseconds

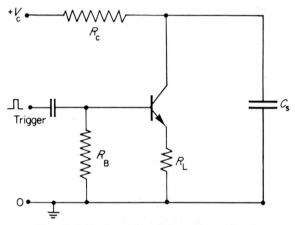

FIG. 7.6. Basic avalanche transistor circuit

for krytrons and may be several milliseconds for thyratrons so the maximum repetition rates are only about 1000 and 100 pulses per second, respectively. Discharge tubes have a limited life and often fail after about 10^7 pulses.

[v] *Avalanche transistors.* The avalanche multiplication associated with the collector breakdown of a silicon transistor often leads to a negative resistance region between emitter and collector. The transistor can be triggered at the base to switch rapidly from a high-impedance to a low-impedance state across this region and for selected transistors the switching time can be less than a nanosecond and quite stable from pulse to pulse.

The basic circuit is shown in Fig. 7.6. The storage condenser C_s is charged to voltage V_c through R_C and then discharged rapidly through R_L by the triggered avalanche of the transistor. R_C must be large enough to allow the collector voltage to fall sufficiently at the end of the pulse to return the transistor to a high-impedance state. R_B should be large enough to limit the collector current before avalanche, but not so large that the effect of the base-collector capacitance becomes appreciable. In practice R_B may be about 50 Ω. C_s must be large enough to provide adequate current through the laser, yet small enough to prevent an unduly slow decay at the end of the pulse.

Avalanche transistor circuits have the advantages of very fast switching speeds at high repetition rates, but the peak current for a single transistor is restricted to less than 20 A by the effective collector spreading resistance in the avalanche state which is usually about 5 Ω. Higher currents can be obtained by using parallel avalanche transistors each with its own storage element provided they can all be triggered simultaneously. Unfortunately variations in device characteristics cause the individual transistors to avalanche at slightly different trigger voltages and the resulting pulse is distorted, with a lower maximum current and a slower rise-time. The problem can be solved by careful transistor selection (Brown, 1966; Karlsons, 1964) or by the provision of adjustable delay lines in the emitter and base leads of each transistor (Prince, 1965). A simpler solution (Hansen, 1967) is to adjust the collector bias voltage of each transistor so that they all switch at the same trigger voltage. In practice this can be done without significantly altering the pulse amplitude or rise-time of the individual circuits. This method of triggering control is very versatile and can also be used to generate rapid trains of pulses with variable spacing by biasing the individual transistors to switch at selected times on a comparatively slowly rising trigger pulse. Alternatively, one transistor may be switched slightly after the others to compensate for any negative transient following the main pulse. These techniques have enabled pulses of up to 50 A to be produced into impedances of 1 Ω with rise-times of 5 nsec at repetition rates of up to 20 KHz. There is no doubt that avalanche transistor circuits have an important part to play in the generation of fast rise-time pulses for gallium arsenide laser systems.

From the above paragraphs it can be concluded that thyristors are the best general-purpose switches if only moderate rise-times are required. Faster pulses can be obtained from mechanical switches and krytrons, but if high repetition rates are needed avalanche transistors must be used.

7.3.4 Pulse transformers

A pulse transformer is often necessary in laser circuits either to increase the current available or to ease the problem of matching into the very low impedance of the laser. The turns ratio is usually less than ten, and values of three to five are frequently used, which means that the impedance of the primary circuit can be an order of magnitude higher than that of the laser circuit. This is very useful when the intrinsic resistance of the switching element is so high that the switch would dissipate more power than the laser. It also means that for a given pulse rise-time a higher value of inductance can be tolerated in the primary circuit, so that the electrical configuration of the switch and the internal inductance in the storage element are less important. Obviously the inductance in the secondary circuit is still critical and it now includes a contribution due to the leakage inductance of the transformer. This arises from stray magnetic flux and is reduced by minimizing the space between the primary and secondary windings. Flux leakage can be further reduced by using a single turn secondary winding which completely encloses the primary turns and the core, and it is often convenient to use the metal case of the transformer as this secondary winding. Ferrite toroidal cores are commonly used since this material has low loss at the high frequencies necessary for pulse reproduction.

Diagrammatic sections of such a transformer are shown in Fig. 7.7. The dimensions depend mainly on the size and shape of the ferrite core. At low duty cycles the cross-sectional area of the core must be large enough to prevent saturation of the magnetic flux during a single pulse and the minimum area required will be proportional to the pulse length and voltage. At high repetition rates heating due to losses in the core and the windings will become more important and it may be necessary to increase the core size and work well below the saturation flux density of the ferrite, since the core losses increase rapidly with increasing flux density. Thin copper strip is used for the primary winding as shown in Fig. 7.7b. This fits closely to the core and also forms a convenient input strip-line which can be matched to the pulse generator. The copper strip and the insulation can each have a thickness of only one or two thousandths of an inch, so that any gaps between the outer case and the core are very small. It is desirable to keep the primary turns evenly distributed around the circumference of the core, and if only a small turns ratio is required it may be better to use two or more primary windings connected in parallel. The ultimate in performance can be obtained by evaporating and plating the secondary metal conductor directly on to the primary insulation, but this is unnecessary in most cases. The output connection from the transformer is shown as a coaxial line in Fig. 7.7a, but it could equally well be a strip-line. The output line should be matched to the impedance of the laser and monitor resistance (if used); but a perfect match cannot be achieved since the laser is a forward biased diode and has an impedance which varies rapidly with voltage. In practice a reasonable match is obtained by using the chord impedance at the working point, and if the laser current is sufficiently high this is also approximately the slope impedance, which is simply the series resistance of the laser. This type of pulse transformer was used in the thyristor-switched laser transmitter described in Section 7.4.7 below, and its shape and size can be judged from the photograph (Fig. 7.12). It is capable of handling 250 A pulses with a rise-time of 30 nsec which is limited by the switching speed of the thyristor.

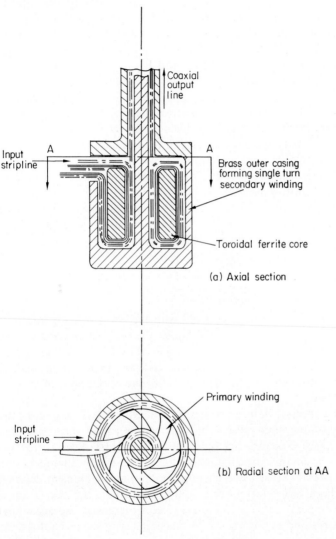

(a) Axial section

(b) Radial section at AA

FIG. 7.7. Diagrams of toroidal pulse transformer: (a) Axial section. (b) Radial section at AA

For low-duty cycles the transformer can be quite small and can be placed close to the laser to reduce the inductance of the connecting leads. This is useful for cooled lasers and transformers have operated successfully at 77°K inside the dewar with the laser, giving output pulses of 50 A with a 10 nsec rise-time at repetition rates up to 10^4 pulses per second.

An alternative type of transformer which is useful for short pulses is the series–parallel transformer (Matick, 1968). This consists of a number of transmission lines of equal length connected in series at one end and in parallel at the other. If there are n lines each of impedance Z, the impedance presented at the series or input end is nZ provided they are correctly terminated at the parallel or output end with an impedance of n/Z. The impedance ratio is n^2 and the structure acts as a pulse transformer of ratio n for pulses shorter than the electrical length of the lines. If n is a fairly small number the transmission lines of the transformer can also serve as efficient leads to the laser which can then be situated some distance away from the pulse generator. Miniature coaxial cable is ideal for this type of transformer, but is difficult to obtain with a suitably low impedance.

Pulse transformers are particularly useful with avalanche transistor circuits where the current is limited by the high internal impedance of the switch. They offer a simple alternative to the use of parallel transistors and eliminate the problem of parallel triggering.

7.4 Cooling

7.4.1 The need for cooling

Chapter 3 dealt in some detail with the effect of temperature on laser parameters. It was seen that although the peak powers and efficiencies of lasers at room temperature are only slightly less than those obtained by cooling, the mean powers are very much lower. In practice a mean power of a few watts can be obtained at 77°K but only about 50 mW at room temperature.

If the ultimate in mean power is required it may be necessary to cool to liquid helium temperatures, but this will inevitably mean a sophisticated and expensive cooling arrangement, and adequate laser performance can usually be obtained by cooling to liquid-air or -nitrogen temperature, which is considerably easier and cheaper. In the past, equipment designers have often been very reluctant to employ such cooling because of unreliability and increased cost and complexity. However, with the accumulation of experience and the steady improvement in cryogenic technology, liquid-air cooling can now be used confidently and conveniently in gallium arsenide laser systems.

There are a variety of ways in which lasers can be cooled and the more important ones are listed below and discussed in turn in the following sections: thermo-electric cooling; mechanical refrigerators; direct cooling by liquids; direct cooling by solids; Joule–Thomson expansion coolers.

The temperature will in general be set by the laser performance required, but the method of cooling will depend on the type, size and cost of the system, the amount of heat to be removed, the power supplies available and the conditions of operation. If full use is made of the cooling to increase the mean output power, the amount of heat to be removed will be in the range of 1 to 10 W for a single laser, and correspondingly more for an array of lasers.

7.4.2 Thermoelectric cooling

At first sight the use of Peltier cooling appears attractive in that only a source of electrical power is required. However, the temperatures that can be reached are very limited, particularly if large amounts of heat have to be removed. Fig. 7.8 shows the temperatures achieved by various types of cooling as a function of heat load, and curves A, B and C indicate the performance of single-stage, two-stage and three-stage Peltier coolers, respectively. Even with three stages in series the lowest attainable temperature is only about $-100°C$ and the actual temperature depends strongly on the heat load so that only small fractions of a watt can be removed before the system warms up considerably. The single-stage cooler can cope with a larger heat load, but the temperatures are higher. Curve A represents a 12 couple 15 A unit which is about 1 in square by $\frac{1}{4}$ in thick and requires about 20 W input power. More heat can be extracted by using much larger units, but the dissipation of the cooler itself goes up proportionately and water-cooling is necessary to remove the heat produced at the hot junction. It is also necessary to ensure stabilization and smoothing of the d.c. supply to the cooler as any ripple will reduce the efficiency considerably, and the original attractively simple system has rapidly become much more complex.

In spite of their limitations Peltier coolers may be useful where only a small improvement on room temperature laser performance is needed. They can also be very useful as supplementary coolers since they can automatically be switched on whenever the ambient temperature rises above a particular value. In this respect it is worth noting that their performance and efficiency improve as the ambient temperature increases.

7.4.3 Mechanical refrigerators

Mechanical heat engines based on a modified Stirling cycle will produce temperatures down to about $20°K$ with no heat load and can remove about

1 W at 30°K or 10 W at 80°K as shown by curve D in Fig. 7.8. They are convenient to operate, requiring only an electrical input of about 500 W and either forced-air or liquid cooling. One obvious disadvantage is that the temperature attained depends on the heat load as well as on the ambient and will fluctuate if either varies. This can be overcome by providing a

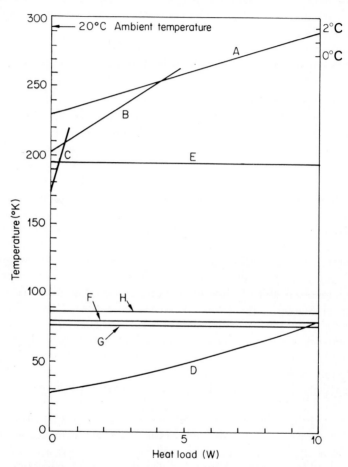

FIG. 7.8. Cold temperature as a function of heat load for various types of cooling: A, Thermoelectric cooler—single stage. B, Thermoelectric cooler—two stage. C, Thermo-electric cooler—three stage. D, Mechanical heat engine (Stirling cycle). E, Solid carbon dioxide. F, Joule–Thomson cooler—air. G, Joule–Thomson cooler—nitrogen. H, Joule–Thomson cooler—argon

controlled heat load and suitable feedback, although the cooling performance of the machine will inevitably be degraded. Heat engines normally use helium as a working fluid and their mechanical specification is very severe as the moving parts have to operate without any lubrication over the whole temperature range from ambient down to 20°K. This means that they are very expensive to manufacture and need inspection and maintenance at relatively short intervals. Provision of pure helium may also be necessary for replenishing the system periodically to replace any gas lost by leakage.

The cooled surface is a metal plate about an inch in diameter and the laser may be mounted directly on this provided suitable attention is paid to the problems of electrical and thermal coupling. Vibration of the cold surface is difficult to suppress and can be a problem. These machines have a considerable thermal mass and take about 10 minutes to reach a stable operating temperature so they are unsuitable for systems requiring a rapid cool-down time. In spite of all these minor disadvantages they have been used extensively with gallium arsenide lasers, particularly for laboratory measurements over a wide range of temperature.

7.4.4 Direct cooling by liquids

The liquefaction of gases and use of the latent heat of evaporation to extract heat from a system has long been a laboratory and industrial practice, and commercial plants are readily available to produce large quantities of liquid air, oxygen, nitrogen, hydrogen and helium at low cost. The direct use of liquids in operational systems raises a variety of problems, but if a supply is available it merits consideration as it offers immediate and continuous cooling to a fixed temperature. For short operating times it can compete favourably on a weight and cost basis with other methods; for example a 1 litre dewar of liquid nitrogen would weigh about 5 lb and would last for $4\frac{1}{2}$ hours with a heat load of 10 W. The same dewar of liquid oxygen would last for almost 7 hours since the latent heat of liquid oxygen is 242 J/cm³, which is 50% higher than that of liquid nitrogen.

The amount of cooling available is in practice limited by the difficulty of transferring the heat from the object to be cooled to the liquid gas. Ruzicka (1958) showed experimentally that the optimum rate of heat extraction from a metal immersed in liquid nitrogen is about 10 W per square centimetre of surface area, and that at this rate the temperature difference across the interface is about 10°K. At higher heat flows a continuous film of gas is formed on the metal surface and much larger temperature differences occur. It is possible to achieve much higher rates of heat transfer by directing a high-velocity jet of liquid on to the metal surface at an appropriate angle, and this approach may be necessary for high power systems requiring a short cool-down time when the thermal mass to be cooled must be minimized. Joule–Thomson coolers (see Section 7.4.6 below) produce a suitable jet of liquid gas and have been used to cool gallium arsenide lasers in this manner. However, if the cool-down time is not critical it is much easier to design the dewar so that a heat-transfer rate of 10 W per square centimetre of interface is not exceeded.

For some applications it may not be convenient to attach the laser directly to a large dewar as this may impose too severe a restriction on the optics. In this case it is possible to transfer the liquid from the dewar to the laser through insulated pipes provided an escape valve is included for venting the gas produced at the laser end of the system. It is also possible to use a novel mode of liquid transfer down flexible polythene pipes in which the liquid is insulated from the inside wall of the pipe by a film of evaporated gas. The pipe remains at room temperature and the heat losses can be very small. Unfortunately this mode of operation depends critically on having a steady liquid flow rate and the correct initial launching conditions, and although it can be achieved reproducibly over appreciable distances in the laboratory it is doubtful whether it is practicable for systems work.

7.4.5   Direct cooling by solids

The latent heat of fusion of solids can be used in a similar way to the latent heat of evaporation of liquids with the same advantage of cooling to a fixed temperature. The solid is usually mixed with a suitable liquid to improve the thermal contact with the object to be cooled. For laser work solid cooling has no advantages over liquid cooling since the temperatures reached by liquefying the common gases are both adequate and convenient. The most common solid coolant is carbon dioxide which sublimes at a temperature of 195°K. Cooling by solid carbon dioxide is represented by the line E in Fig. 7.8.

7.4.6   Joule–Thomson expansion coolers

These coolers could have been included in the section on liquid cooling, but they have proved so successful for cooling gallium arsenide lasers and other semiconductor devices that a separate treatment is justified. The basic idea is to produce the liquid used for cooling inside the system itself and a miniature Joule–Thomson expansion cooler operating from a high pressure supply of gas is the most convenient liquefier for this purpose. Provided the gas has an inversion temperature above room temperature it will cool on expansion through an orifice. For air expanding from

3000 p.s.i. to atmospheric pressure the temperature drop is about 37°K. This temperature drop can be transferred to the incoming high pressure gas by an efficient heat exchanger, and when this cooled air expands its temperature is reduced still further. The process continues until the gas leaving the orifice is liquefied. The heat exchanger and the expansion orifice or nozzle (which is typically one or two thousandths of an inch in diameter) are the two basic components of the liquefier. The heat exchanger

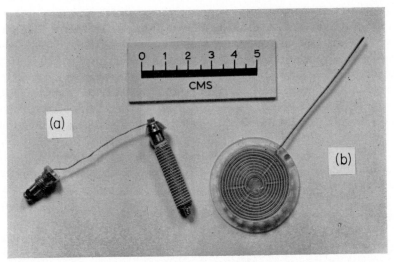

Fig. 7.9. Joule–Thomson expansion liquefiers: (a) 'Minicooler' (Type MC8). (b) 'Emicooler'

is made from small-bore finned tubing wound into either a spiral or a helical coil. The high pressure gas passes down the bore and the cold expanded gas is constrained to pass through the fins. These heat exchangers can be up to 95% efficient and the exhaust gas will be only a few degrees below ambient temperature. Fig. 7.9 shows photographs of two types of liquefier; (a) is a "Minicooler" (Type MC8) made by the Hymatic Engineering Co. Ltd., Redditch, Worcestershire, and (b) is an "Emicooler" made by E.M.I., Feltham, Middlesex. It is obvious that the very small size of these coolers makes them ideal for systems work.

Most of the common gases with the exception of hydrogen and helium can be liquefied directly by this method. Those most widely used for laser cooling are air, nitrogen and argon giving temperatures of 80, 77·4 and 90·2°K, respectively. Argon has the largest enthalpy change and a low specific heat; for a given volume of gas it produces about twice as much cooling as air or nitrogen. It is especially useful where very fast cool-down times are required, and the "Emicooler" shown in Fig. 7.9b (which was designed with a very small thermal mass to give rapid cooling) will produce liquid argon from room temperature in less than 1 second.

The temperature reached is, of course, the boiling point of the gas used and is constant for all heat loads up to the limit of the cooler. This limit depends on the size of orifice, the gas pressure and the ambient temperature, being approximately proportional to the first two, and falling by about 10% for every 10 degree rise in ambient temperature above 20°C. About 10 W of cooling can be produced efficiently at 20°C ambient temperature using air or nitrogen at 3000 p.s.i., and about 20 W using argon. These performances are represented in Fig. 7.8 by the lines F, G and H, respectively. Cool-down times of less than 1 minute are typical, but can be less than 1 second if required, although such rapid cooling is only achieved with very low heat loads. To make full use of this speed the laser mount must be carefully designed as an integral part of the cooler to give good heat transfer with low thermal mass.

Joule–Thomson coolers have been used for several years, mainly for cooling infrared photodetectors to reduce the noise level. In this respect they suffered from three grave disadvantages.

[1] Microphony. The detector performance was often limited by low-frequency noise produced by the bubbling of the boiling liquid.

[2] The need for extremely clean pure gas. Any small particles in the gas or any impurities which freeze out at the liquid temperature would readily block the tiny hole in the expansion nozzle.

[3] The size and weight of the gas supply. The use of high-pressure gas cylinders could only be considered for short duration systems, and if a compressor and filters were used the filters had to be changed or regenerated frequently to maintain an adequate standard of gas cleanliness.

However, for cooling gallium arsenide lasers different conditions apply and the disadvantages are largely eliminated for the following reasons.

[1] Microphony is no longer a problem since the laser is a high-power transmitter and not a low power receiver so the effect of the vibration is negligible.

[2] The laser produces much more heat than did the photodetectors. Consequently, much more cooling is required, and the easiest way to obtain this is to increase the gas flow by enlarging the expansion orifice. Larger solid particles can be tolerated in the gas and it is more difficult for the hole to be blocked by frozen impurities against the increased force of the gas flow. This simplifies the problem of gas cleaning and filtration, reducing the cost and complexity of the system and virtually eliminating

the risk of blockage, especially on systems with a limited period of operation. In the rare event of blockage occurring it is usually caused by freezing of impurities in the nozzle, and simply allowing the system to warm up towards room temperature will remove the cause of the trouble.

[3] Since the lasers need a higher gas flow than did the detectors, it would seem that the problem of bulky gas supplies must be more serious. However, a recent breakthrough in liquefier technology has improved things considerably by enabling the gas flow at the nozzle to be controlled to give the exact amount of cooling required at any instant, thus avoiding wastage of gas. The maximum cooling power and hence the highest gas flow needed for a particular system will be determined either by the steady-state demand at the highest operating ambient temperature or by the necessity to cool the laser down within a specified time. The nozzle size and gas pressure must be chosen to meet this maximum demand with the result that for most of the time the cooler is producing liquid at a higher rate than necessary. The excess liquid floods the heat exchanger and reduces its efficiency until a dynamic equilibrium is reached which automatically adjusts to changes of ambient temperature or heat load. The gas flow can be matched to the cooling need at any time by controlling the input pressure, but this is an inefficient solution since a large amount of gas enthalpy is lost in the pressure drop through the control valve. It is much better to restrict the flow at the actual expansion nozzle so that all the pressure drop occurs there and no enthalpy is wasted. This must be done by changing the size of the orifice in accordance with the thermal load. Such a cooler has recently been developed by the Hymatic Engineering Co. Ltd. Any variation in liquid level in the reservoir automatically alters the size of the orifice and the gas flow either increases or decreases until the original level is restored. The maximum gas flow is available for the initial cool down and no pressure reducing valves are needed, so the system is reduced to a minimum number of components and can be appreciably smaller and lighter in weight than with a conventional cooler as well as being more efficient.

The amount of gas saved by the self-regulating cooler will depend on the exact conditions of operation but for a continuous heat load of 1 W the average gas consumption should be reduced by a factor of four, and if long periods of standby at low heat loads are needed the benefit can obviously be much greater.

A further improvement can be obtained by using supplementary cooling to cope with the highest ambient temperatures where the performance of the expansion cooler deteriorates rapidly, and "Minicoolers" are available with an integral precooling stage which uses the expansion of Freon. The flow of Freon is controlled according to the ambient temperature and this gives a very economical system. Again it is rather arbitrary to give quantitative figures without specifying details of the system, but another reduction of four times in the average gas consumption can often be achieved.

An idea of the overall size and weight of Joule–Thomson cooling systems can best be obtained by considering a particular requirement. For a single gallium arsenide laser operating at 50 W peak power and a

TABLE 7.1. Size and weight of cooling systems for 12-hour operation with 1 watt heat load.

	Standard Minicooler system	Self-regulating Minicooler system	Precooled self-regulating system
Bottle capacity (litres)	92·5	24·0	5·0 (+0·6 Freon)
Bottle weight (kg)	134·0	35·5	6·4 (+0·9 Freon)
System weight (kg)	170·0	45·5	13·6

relatively low duty cycle of 10^{-3} about 1 W of cooling is needed. Assuming that it must operate continuously for 12 hours from gas bottles charged to 4000 p.s.i. in an ambient temperature ranging from -20 to $+60°C$ (which is a much larger variation than is likely to be met in practice), the bottle capacity and the total weight of the cooling system can be estimated for an ordinary Minicooler, a self-regulating Minicooler and a self-regulating Minicooler with Freon precooling. The results are given in Table 7.1 and the advantage of the self-regulating and precooled systems can be clearly seen. It should be noted that the extra size and weight necessary for the Freon precooling (given in brackets in Table 7.1) are only a small fraction of the total, and it is certainly worth tolerating the extra complexity involved. Extrapolations from this example can be made, but if the requirement is vastly different it will be better to repeat the basic calculation for the system considered.

Storage of clean gas in high pressure cylinders is convenient for low powers and limited periods of operation, but for higher power transmitters operating for indefinite periods it is uneconomic, and it is better to use a compressor to supply the high pressure gas. One or more stages of compression can be used and the system can be either recirculating or open.

A recirculating system collects the clean exhaust gas from the cooler and recompresses it. With a perfectly clean compressor no gas cleaning would be necessary, but in practice a molecular sieve must be used to remove hydrocarbons and other impurities introduced during compression. Obviously, dry compressors will introduce less contamination than oil-lubricated ones, but usually need more frequent maintenance. Diaphragm compressors can approach ideal cleanliness, but the life of the diaphragm is limited and it must be changed regularly to avoid the risk of rupture and gross contamination of the whole system. Recirculating systems can be used equally well with air, nitrogen or argon, but the provision of leak-free collection of the exhaust gas can sometimes be a minor inconvenience.

Open systems, on the other hand, can only be used with air, which is taken in from the atmosphere. Since the air needs filtering and cleaning thoroughly before it can be used there is little point in using a clean compressor since the filters will also take out any compressor contamination. The usual sequence of filtration is to use a gross oil and water trap followed by soda lime to remove the carbon dioxide, activated charcoal to remove hydrocarbons and a molecular sieve to remove water vapour and any traces of contaminants which may pass through or be carried over from the other absorbents. Although quite rigorous cleaning standards are required the filter units need not be very bulky, and three filter vessels of only 1·6 litres capacity each can clean enough air to run one cooler for 300 hours.

Fig. 7.10 is a photograph of a clean air supply plant designed for the laboratory use of Joule–Thomson coolers, and a functional block diagram is shown in Fig. 7.11. Atmospheric air is compressed to 3000 p.s.i., filtered and stored for use at this pressure. A booster unit provides subsequent compression up to 8000 p.s.i. as required.

There is no doubt that Joule–Thomson expansion coolers offer a very efficient and versatile means of cooling gallium arsenide lasers to a stable low temperature, and it is the author's opinion that they will be used in a large number of systems on the basis of their low cost, ruggedness and simplicity.

7.4.7 *Design of a compact cooled transmitter*

It is instructive to conclude this section on cooling by considering some of the thermal problems encountered in the design of a small general-purpose laser transmitter to operate with a single high-power gallium arsenide laser at liquid-air temperature. Fig. 7.12 is a photograph of such a transmitter described by Hambleton (1964).

One of the most important thermal problems in a high mean power transmitter is the design of the electrical leads since they inevitably account for a considerable proportion of the total heat loss. It can easily be shown that practically all the heat generated electrically in the leads flows out at the low-temperature end, so the total heat flow down the leads is the sum

Fig. 7.10. Clean air supply plant for Joule–Thomson coolers

of this Joule heating and the thermal conduction due to the ends being at different temperatures. For each conductor the total heat flow is:

$$H = \frac{I^2 \rho l \tau}{A} + \frac{K A \Delta T}{l} \quad \text{W}, \tag{7.8}$$

where ρ is the average electrical resistivity, K is the average thermal conductivity, l is the length of the conductor, A is the cross-sectional area of the conductor, ΔT is the temperature difference down the conductor, I is the pulse current and τ is the duty cycle.

For a given conductor material the cross-sectional area must be optimized since increasing it will reduce the Joule heating but increase the thermal conduction, and the minimum total loss will occur when $dH/dA = 0$. This is obviously a minimum condition since $H = \infty$ for either $A = 0$ or $A = \infty$.

Applications of Gallium Arsenide Lasers

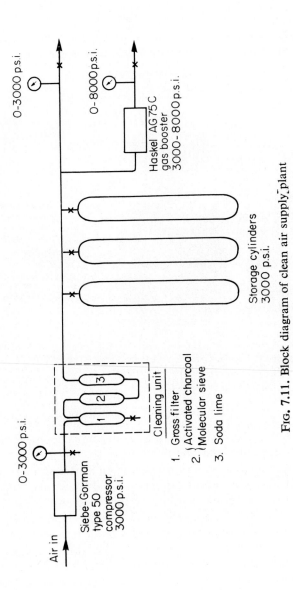

FIG. 7.11. Block diagram of clean air supply plant

Hence the optimum area is given by

$$A = Il\sqrt{\frac{\rho\tau}{K\Delta T}}.$$ (7.9)

If this optimum area is used the Joule heating term is equal to the thermal conduction term and the total heat flow is given by

$$H = 2I\sqrt{\rho K\tau\Delta T}.$$ (7.10)

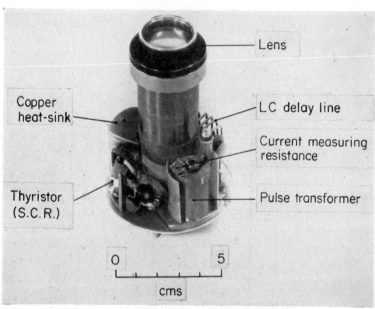

FIG. 7.12. Compact 100 W laser transmitter

In Eqn. (7.8) ρ and K were taken as average quantities between 80°K and room temperature and this is adequate for the present discussion, although if accurate calculations are required it is necessary to replace the terms ρl and $K\Delta T/l$ by the appropriate integrals down the length of the conductor. Two significant conclusions can be drawn from the above equations. Firstly, the conductor length l does not appear in Eqn. (7.10). Hence a short thermal system can be used without increasing the total heat loss provided the cross-sectional area of the electrical conductors is reduced proportionately. This is obviously important in the design of small transmitters. Keeping the length short will also reduce the surface area which is below room temperature, and this helps to reduce the thermal loss due to radiation which can be taken as 0·046 W/cm² at 80°K.

Secondly, Eqn. (7.10) shows that, since I, τ and ΔT are fixed by the operating conditions, the choice of metal for the conductors only affects the total heat loss through the product ρK. But this product is approximately constant for most metals as implied by the Wiedemann–Franz law, so the heat loss is almost independent of the material used. Hence the choice of metal can be governed by more practical factors and in the transmitter shown in Fig. 7.12 Nilo K was used. The optimum cross-sectional area was much larger for this than for a good electrical conductor such as copper, so the conductors were mechanically stronger. With Nilo K it was also much easier to make a good glass-to-metal vacuum seal between the conductors.

Another important problem in the design of a cooled laser transmitter is the provision of thermal insulation. A vacuum is the obvious choice since this provides a suitably clean environment for the laser in addition to thermal insulation. However, a final outgassing at high temperature cannot be carried out without damaging the laser, and this means that adsorbed gas on the dewar surfaces cannot be effectively removed and is gradually released after the system is evacuated and sealed off.

The simplest solution is to take the laser out of the vacuum space and put it inside the inner wall of the dewar which can now be thoroughly baked and outgassed to hold a permanent vacuum. The laser radiation must pass through two glass windows instead of only one but the loss of power is small especially if bloomed windows are used. If a Joule–Thomson expansion cooler is used a clean ambient for the laser can be easily and conveniently provided by allowing the cold clean air inside the cooler to fill the laser compartment. If a little extra heat loss can be tolerated the vacuum can be dispensed with altogether and expanded polystyrene used for thermal insulation. Condensation and frosting on the outer window or lens surface can be prevented by constraining the exhaust clean air from the Joule–Thomson cooler to flow past this exposed surface so that no moisture reaches it.

It is appropriate to end this section on cooling by summarizing the major heat losses in a simple small transmitter. The following data refer to the particular transmitter shown in Fig. 7.12, when operating at 100 W peak and 0·5 W mean output power.

Electrical dissipation at the laser junction	1 W
Electrical dissipation in the laser series resistance	3 W
Electrical dissipation in the conducting leads	2 W
Thermal conduction down the leads	2 W
Radiation losses	1 W
Total heat losses	9 W

With the laser switched off, the heat loss on standby was reduced to 3 W. The overall thermal efficiency was about 5%, and the above figures emphasize the importance of correct electrical lead design for high mean power operation, since the leads account for almost half of the total thermal loss.

7.5 Optics

7.5.1 *Spherical lenses*

For practical applications a gallium arsenide laser is best considered as a line source of infrared radiation with dimensions about 1 mm by 2 μ. The coherence can usually be ignored, especially for pulsed operation, and the maximum optical power density at the emitting aperture is about 5×10^6 W/cm^2 with a corresponding brightness of about 2×10^7 W/cm^2 steradian. Essentially it is the brightness that is the important property for systems work, since an increase in source brightness can mean either a better performance with the same optical system or a smaller and simpler optical system for the same performance.

The function of the transmitter lens or lenses is to transform the beamwidth and aperture of the laser itself to the optimum values for a particular application, and the first criterion to establish is the lens aperture required to collect all or nearly all of the laser radiation. Measurements of the angular width of emission in planes parallel and perpendicular to the junction are described in Chapter 3. At high powers there is little difference in the divergence in these two planes. In practice it is found that an $f/2$ spherical lens will collect at least 90% of the radiation whereas an $f/3$ lens may collect only 60 to 70%.

The choice of lens focal length will be governed by the beamwidth requirements of the particular system considered. If a simple spherical lens is used the beamwidth will be unequal in the two dimensions. In theory, a 10 cm focal length lens used with a 1 mm wide laser would give beamwidths of 10 milliradians and 0·02 milliradians in azimuth and elevation, respectively. In practice, the narrow beamwidth will not be determined by the focal length, but by the mechanical difficulty of positioning the emitting area of the laser exactly at the focus of the lens, and indeed the laser may be defocused deliberately to make the beamwidths in the two dimensions more comparable in magnitude.

7.5.2 *Cylindrical lenses*

A more elegant way of obtaining equal beamwidths and at the same time reducing the size of the optics is to use two cylindrical lenses instead of one spherical one. The principle is made clear in Fig. 7.13 where the diagrams

(a) and (b) show sections of the optics in azimuth and elevation, respectively. Lens A is a cylindrical lens with a short focal length f_1. It collimates the light in elevation only and the beamwidth is given by L_1/f_1, where L_1 is the shorter dimension of the emitting aperture. Lens B is a cylindrical lens at right angles with a focal length f_2 which collimates the light in azimuth

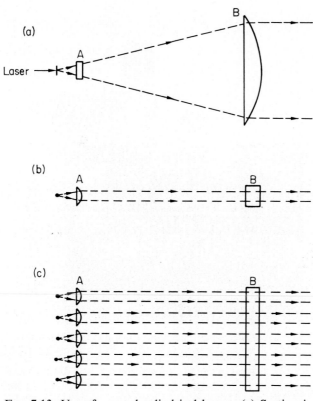

Fig. 7.13. Use of crossed cylindrical lenses: (a) Section in azimuth. (b) Section in elevation. (c) Extension to a linear array

to a beamwidth of L_2/f_2, where L_2 is the longer dimension of the emitting aperture. Obviously, the beamwidths can be made equal by choosing $f_1/f_2 = L_1/L_2$. It is clear that the output aperture and hence the total size and weight of the optics is reduced considerably. For example, if a maximum beamwidth of 1 milliradian by 1 milliradian were required from a 1 mm wide laser, the use of $f/2$ cylindrical lenses would result in an output aperture of 50 cm by possibly 0·5 cm, whereas with an $f/2$ spherical lens the aperture would be 50 cm diameter, and the volume occupied by the optics would be very large.

Reducing the size of the optics for a given performance is particularly important when arrays of lasers are considered. Fig. 7.13c shows a simple extension of the use of cylindrical optics to a line of lasers, where the lens B

Fig. 7.14. Multilaser transmitter using cylindrical optics

is common to all lasers. In the more general case two-dimensional stacking could be used with the A lenses common to whole rows and the B lenses common to whole columns so that a rectangular array of $m \times n$ lasers could be focused using only $m + n$ lenses.

A photograph of a transmitter using cylindrical optics with a single line of ten lasers is shown in Fig. 7.14. The lasers are connected electrically in series to increase the impedance and simplify the current drive requirements. They are mounted on an anodized aluminium plate which gives electrical insulation with good thermal contact and are cooled by liquid nitrogen. This transmitter is capable of producing a peak power of 500 W and a mean power of about 5 W. The output beamwidth is 5 mradian in azimuth limited by the focal length of the output lens and 2 milliradian in elevation limited by the mechanical accuracy with which the ten small lenses can be positioned. The output aperture is 7 cm by 8 cm. Hence the brightness of the output beam is about 10^6 W/cm² steradian which is only a factor of ten below the brightness of a single 50 W laser. It is apparent that with careful attention to mechanical design arrays of cylindrical lenses can be used very effectively, and efficient stacking of lasers can be achieved.

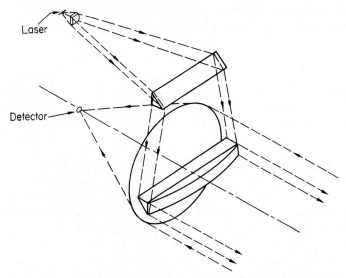

FIG. 7.15. Coaxial optical system using cylindrical transmitter lenses

The use of cylindrical lenses to generate a narrow beamwidth in both azimuth and elevation from a single gallium arsenide laser results in a long narrow output aperture. In most systems the receiving aperture is circular and as large as possible to collect the maximum amount of power. These two apertures can often be combined by using prisms or plane mirrors so that the transmitting aperture lies along a diameter of the receiving one, resulting in a very compact concentric optical system. The simplest arrangement is shown in Fig. 7.15, and it can be easily adapted to a variety of configurations. The success of this approach relies on the fact that the transmitting aperture can be very narrow and need only obscure a small fraction of the receiving lens or mirror. Assuming that the small dimension of the transmitting aperture is 0·5 cm (which is large enough to produce a beamwidth as small as 1 milliradian) and the large dimension is equal to the diameter of the receiving aperture, only about 2% of a 30 cm aperture is obscured.

7.5.3   Fibre optics

Spherical or cylindrical lenses are the simplest and most common means of changing the beamwidth and aperture of the laser to the values required by a particular system, but the use of fibre optics or cone channel optics can also be considered. In recent years the technology of fibre optics has been developed to a high level of sophistication, and fibre diameters ranging from micrometres to millimetres are readily available both as individual fibres or as coherent bundles where each fibre occupies the same relative position at the two ends.

The properties and use of such fibres have been described by several authors (see, for instance, Kapany and Capellaro, 1961). Basically, a fibre of refractive index N_1 coated with a thin protective sheath of refractive index N_2 will transmit without loss, other than that due to absorption, light which is incident on the end face at angles up to θ_0 given by

$$N_0 \sin \theta_0 = (N_1^2 - N_2^2)^{\frac{1}{2}} \tag{7.11}$$

N_0 is the refractive index of the medium in which the fibre is immersed and the quantity $N_0 \sin \theta_0$ is called the numerical aperture of the fibre. Light incident at angles greater than θ_0 does not undergo total internal reflection at the interface between the core and the protective sheath, and is soon lost from the fibre by scattering at surface imperfections. The types of glass in common use give numerical apertures up to about 0·7 corresponding to values of θ_0 of up to 60° in air. Hence coated fibres are easily capable of transmitting all the radiation emitted by a gallium arsenide laser, and since the fibre diameters and apertures can be comparable in size to the laser aperture it is logical to consider them as optical components in laser systems.

The simplest use of fibre optics is to convert the line source of laser radiation into an equivalent circular area so that spherical lenses can be used efficiently, and this has been tried in several laboratories. The main problems are to polish the ends of the small fibre bundles optically flat, and to position the input aperture sufficiently accurately with respect to the laser.

Suitably shaped fibre bundles have been fabricated and polished to give transmission factors of up to 50% for light at near normal incidence. The degree of perfection is limited by the different polishing rates of the fibres and the resin in which they are normally held, resulting in rounding and distortion of the ends of the individual fibres. It may be possible to avoid this effect by fusing the fibre bundle into a glass matrix rather than enclosing the fibres in resin, so that the whole assembly would polish at a constant rate.

The choice of fibre diameter is governed by the mechanical precision with which the bundle can be held and aligned, and in practice the best results have been obtained with fibre diameters between 20 and 40 μ. With a 1 mm laser junction the input end of the bundle will consist of a single straight line of 25 to 50 contiguous fibres, and the output end will have the fibres hexagonally close packed into a circle of diameter about 150 to 200 μ. If 20 μ diameter fibres are used the input end of the bundle must be placed within 40 μ of the gallium arsenide surface to give $f/2$ collection of the laser light, and a lateral positional accuracy of a few micrometres is necessary in the direction perpendicular to the laser junction. Such precision is difficult to achieve at room temperature and virtually impossible with cooled lasers because of differential thermal expansions. Attempts have been made to attach the fibres permanently to the laser but they have not been successful. The positional requirements are less severe for larger fibres, but the output aperture is correspondingly increased and larger lenses will be necessary for a given beamwidth.

In general for a single laser a much better performance will be obtained with cylindrical lenses than with fibre optics, but the latter may be necessary if the use of spherical lenses is unavoidable as, for instance, when the transmitter and receiver must both use the same optical system.

It has also been suggested that fibre optics can be used to combine the beams from several lasers into a single output aperture (Chatterton, 1965). In this case relatively large-diameter fibres may be used to simplify the alignment problems and the output aperture will still be much smaller than the area occupied by the initial array of lasing junctions. The optics may not be as efficient as the cylindrical lens array described in Section 7.5.2 above, but the flexibility of the fibre bundles enables the lasers to be mounted in practically any configuration, and this may be important if large numbers are to be used.

7.5.4   Cone channel optics

In the previous section on fibre optics it was assumed that the fibres were parallel-sided so that the output and input ends of the fibre bundle had the same area. It is also possible to use tapered fibres so that the area as well as the shape of the aperture can be varied. As the fibre radius is reduced from R_1 to R_2 the angles θ_1 and θ_2 of the incident and emergent rays are related by the sine condition of ideal optics, namely:

$$\frac{\sin \theta_1}{\sin \theta_2} = \frac{R_2}{R_1}. \tag{7.12}$$

This was derived for axial rays by Williamson (1952) who carried out a theoretical and experimental study of reflecting cones. Witte (1965) extended the geometric analysis to skew rays and also considered the positions of the entrance and exit pupils of the system. He concluded that the reflecting cone has the properties of an ideal optical system and is often much simpler and more convenient to use than an equivalent combination of lenses, especially if large apertures are required. The single cones considered by Wilkinson and Witte were bounded by reflecting mirrors and had no imaging properties, but composite cones consisting of coherent bundles of tapered fibres are capable of forming images with a resolution equal to the fibre diameter. If coated fibres are used and the angle of taper is small the beam angle at the smaller end is limited to a value less than the critical angle for total internal reflection given by Eqn. (7.11) in the previous section (Kapany, 1961).

Although Eqn. (7.12) implies that the brightness of the beam traversing the cone will be unchanged, cone channel optics can sometimes be used to reduce the total space occupied by the optical system. For use with gallium arsenide lasers they suffer the same serious disadvantage as fibre optics in that it is difficult to couple efficiently to the very small aperture of the laser. They may, however, find use in the detector end of the system where the device areas are appreciably larger. In this respect the inability to form an image may be an advantage since any non-uniformity in response over the area of the detector will be averaged out. In some detector systems large beamwidths are required, and if the detector area is limited this implies the use of a large diameter lens or mirror with a small focal length. Fig. 7.16 illustrates how the performance of a large-aperture lens can be obtained by using a small-aperture lens and a reflecting cone, the focal length being effectively shortened by a factor equal to the cone reduction. The extent to which this technique can be employed depends on the angle of the extreme rays which the detector will accept. It can be used successfully with silicon photodiodes, and with a glass cone the best results will be obtained by bonding the smaller face of the cone directly to the silicon surface with a transparent cement of suitable refractive index. Using cone channel optics it is possible to achieve a performance equivalent to $f/0.7$ from a lens with an aperture of only $f/2$ or $f/3$.

7.6 Detectors

7.6.1 Limitations of detector performances

[i] General definitions. A perfect photodetector would have two fundamental properties. It would be capable of detecting every photon incident upon it within a chosen range of optical frequency, and it would

contribute no noise to the external circuit. A practical device can sometimes approach ideal behaviour as regards either one of these properties, but if so it is usually very bad as far as the other is concerned. The two classes of detector commonly used with gallium arsenide lasers are vacuum phototubes and silicon *p–n* junction detectors. Under certain operating conditions the noise contributed by the vacuum tubes can be extremely small but they depend on the use of photoemissive cathodes which

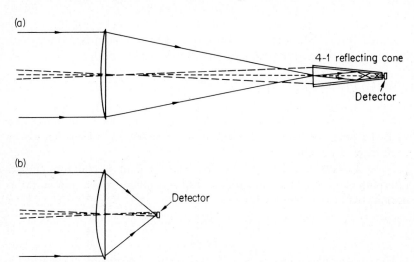

(a)

4-1 reflecting cone

Detector

(b)

Detector

FIG. 7.16. Use of cone channel optics to give an increased aperture: (a) Small aperture lens and reflecting cone. (b) Equivalent large aperture lens

invariably have a very low quantum efficiency, so that only a very small fraction of the incident photons are detected. On the other hand, silicon photodiodes can have quantum efficiencies greater than 50% so that more than half of the incident photons produce electrons, but in general they are much noisier than the phototubes because they have much higher leakage currents.

In comparing the performances of different types of photodector the following terms and definitions are frequently used.

SENSITIVITY, S_0, is the output signal current per watt of radiation incident on the detector.

$$S_0 = \frac{I_s}{P_s} = \frac{\eta e}{h\nu} \quad \text{A/W} \tag{7.13}$$

where I_s is the photocurrent produced by the signal, P_s is the total signal power incident on the detector, η is the quantum efficiency of the detector, e is the electronic charge and $h\nu$ is the energy of each incident photon.

Equation (7.13) assumes that there is no amplification associated with the detection process so that the output current is simply the photoelectron current I_s. If there is gain in the detector and the current is multiplied by a factor m, the sensitivity becomes

$$S_m = \frac{mI_s}{P_s} = \frac{m\eta e}{h\nu} \quad \text{A/W.} \tag{7.14}$$

DETECTIVITY, D^*, is the signal-to-noise ratio produced by unit radiant flux normalized to unit area and unit electrical bandwidth.

$$D^* = \frac{S_0\sqrt{A\Delta f}}{I_n} \quad \text{cm Hz}^{\frac{1}{2}}/\text{W}, \tag{7.15}$$

where I_n is the total noise current in the detector, A is the detector area and Δf is the electrical bandwidth used to measure D^*, and ideally should be as narrow as possible.

NOISE EQUIVALENT POWER, N.E.P., is the minimum radiation power necessary to give a signal-to-noise ratio of unity when the noise is normalized to unit bandwidth.

$$\text{N.E.P.} = \frac{I_n}{S_0\sqrt{\Delta f}} \quad \text{W/Hz}^{\frac{1}{2}}. \tag{7.16}$$

The above definitions for N.E.P. and D^* were originally conceived for work with very narrow bandwidths centred at the various mechanical chopping frequencies commonly used in infrared systems. In normalizing to unit bandwidth it was assumed that the measurement bandwidth, Δf, was centred on the quoted chopping frequency and was sufficiently narrow that the noise power per cycle was constant within Δf. In principle the centre frequency and the measurement bandwidth should always be specified in quoting values of N.E.P. or D^* for a detector. In practice the above definitions will be valid over very large bandwidths provided that the noise voltage or current in the detector is proportional to the square root of the bandwidth. This is true for many but not all types of detector noise, and a more general comparison of detectors can be made by using the total noise-equivalent signal power in a given bandwidth without normalizing to unit bandwidth. In this chapter this will be referred to as the TOTAL NOISE EQUIVALENT POWER, T.N.E.P., and defined as the minimum radiation power in watts necessary to give a signal-to-noise ratio of unity for a specified electrical bandwidth B. The symbol B will be used rather than Δf to imply that the definition remains valid for very large bandwidths.

Unless otherwise stated B will imply a bandwidth extending from 0 to B Hz.

$$\text{T.N.E.P.} = \frac{I_n}{S_0} = \frac{h\nu I_n}{\eta e}. \tag{7.17}$$

There is often confusion regarding N.E.P., and it is sometimes used mistakenly as though it were the T.N.E.P. defined above. The T.N.E.P. gives a direct indication of the minimum power that can be detected in a given bandwidth regardless of the type of noise present. In fact a T.N.E.P. can be quoted for the whole detection system including optical filters and amplifiers, and a graph of T.N.E.P. as a function of bandwidth will characterize the detection system completely, enabling comparisons to be made quickly and easily.

For a given detector the above quantities are all functions of the wavelength, λ, and if numerical values are quoted without specifying λ it is assumed that they are the values at the peak of the spectral response for that particular detector. However, this chapter is solely concerned with gallium arsenide laser wavelengths of 8500 to 9000 Å, and where figures are given they refer to 9000 Å unless otherwise stated.

There are several sources of noise than can limit the performance of a detector, and these will be considered in the following order.

[1] *Thermal noise* in the detector load resistance. This contributes an output noise power which is given by

$$(I_n^2 R)_{\text{thermal}} = 4kTB, \tag{7.18}$$

where R is the load resistance, k is Boltzmann's constant and T is the effective noise temperature of the load resistance.

[2] *Shot noise* on the total detector current, I_t. The total shot noise power is given by

$$(I_n^2 R)_{\text{shot}} = 2eBRI_t. \tag{7.19}$$

The detector current I_t is the sum of the current I_b due to background radiation, the detector dark current I_d and the signal current I_s.

[3] *Amplifier noise*, which depends to some extent on the bandwidth and gain required from the first stage of amplification.

If gain is provided in the detector itself the noise in the following amplifier will be less important, and if the detector currents are multiplied by a factor m the general expression for the signal-to-noise power ratio at the output of the detector is given by

$$\frac{S}{N} = \frac{m^2 I_s^2 R}{2eBRm^2(I_b + I_d + I_s) + 4kTB}. \tag{7.20}$$

The signal current I_s is related to the incident radiation power P_s by Eqn. (7.13) and combining this with Eqn. (7.20) enables the T.N.E.P. to be calculated. In the general case

$$\text{T.N.E.P.} = \frac{h\nu}{\eta e}\left[2eB(I_b + I_d + I_s) + \frac{1}{m^2}\frac{4kTB}{R}\right]^{\frac{1}{2}}. \qquad (7.21)$$

[ii] *Thermal noise.* From Eqns. (7.20) and (7.21) it can be seen that if $2kT > eRm^2 I_t$ the detector performance is limited by the thermal noise in the load resistance. A better performance can be obtained by increasing either R or m until the shot noise dominates. Increasing R is the simpler solution, but this will restrict the bandwidth because of the time constant associated with R and the detector capacitance C. The best approach is to make C as small as possible and then choose R to give the required bandwidth according to the relation $2\pi RCB = 1$. In this case Eqn. (7.21) becomes

$$\text{T.N.E.P.} = \frac{h\nu}{\eta e}\left[2eB(I_b + I_d + I_s) + \frac{B^2}{m^2}8\pi kTC\right]^{\frac{1}{2}} \qquad (7.22)$$

$$= \frac{h\nu}{\eta e}\left[(I_n)^2_{\text{shot}} + \frac{1}{m^2}(I_n)^2_{\text{thermal}}\right]^{\frac{1}{2}}.$$

This equation shows that if R is optimized for any given bandwidth the thermal noise current is directly proportional to bandwidth whereas the shot noise current is proportional to the square root of bandwidth. Hence thermal noise will be important for work with fast pulses, especially if there is little or no gain in the detector. It is obviously bad practice to use D^* to specify the performance if the detector is liable to be limited by thermal noise, since D^* implies the wrong variation of noise with bandwidth. It is better to state the T.N.E.P. either for a given bandwidth or as a function of bandwidth.

Thermal noise can be effectively reduced by increasing the internal current gain in the detector, provided of course that there is no extra noise introduced by the multiplication process. Photomultipliers are particularly good in this respect, and the gain can usually be made large enough for thermal noise to be neglected in photomultiplier systems. Semiconductor avalanche photodiodes can also show high values of current gain although in this case there is an additional noise mechanism associated with the avalanche multiplication process (McIntyre, 1966). Nevertheless, they have a much better performance than ordinary silicon photodiodes at the higher bandwidths where the latter are limited by thermal noise (see Section 7.6.2[v] below).

[iii] *Shot noise due to background radiation.* The detector current due to the background radiation is given by

$$I_{\rm b} = \frac{\eta e P_{\rm n}}{h\nu},$$ (7.23)

where $P_{\rm n}$ is the background power incident on the detector area. For a system in which the largest noise contribution is due to the background, Eqn. (7.20) becomes

$$\frac{S}{N} = \frac{I_{\rm s}^2}{2eBI_{\rm b}} = \frac{\eta P_{\rm s}^2}{2h\nu BP_{\rm n}}.$$ (7.24)

Hence under background-limited conditions the only relevant detector parameter is the quantum efficiency, η, and at 9000 Å a silicon photodiode will give up to 200 times more signal-to-noise power than a photomultiplier. Increasing the collecting area of the receiver will increase $P_{\rm s}$ and $P_{\rm n}$ in the same ratio but Eqn. (7.24) shows that a larger area will give a better performance because the signal power increases more rapidly than the shot noise power.

The background power incident on the detector will depend on the brightness or radiant emittance of the background seen by the detector optics, and is given by

$$P_{\rm n} = MA_{\rm r}\Omega_{\rm r}\Delta\nu,$$ (7.25)

where M is the radiant emittance of the background, $A_{\rm r}$ is the receiver area, $\Omega_{\rm r}$ is the receiver beamwidth and $\Delta\nu$ is the spectral width accepted by the detector system.

For blue sky an average value for M is about 1 $\mu W/cm^2$ steradian Å, but it may be up to an order of magnitude higher or lower than this depending on the position of the sun and the amount and type of cloud (Bolle, 1965). For moonlight the figure is about a million times less, and for starlight M can be as low as 10^{-15} W/cm^2 steradian Å (Biernson, 1963). These figures are also relevant for systems such as laser altimeters which only view the ground since the diffuse reflectance of some types of terrain can approach unity (Ashburn, 1956).

The noise due to background radiation can be reduced by decreasing the angle of view, $\Omega_{\rm r}$, of the detector optics and by using a narrow band optical filter which only transmits radiation at the laser wavelength. Although the spectral width of the laser is only 20 or 30 Å it is usual to make the filter bandwidth about 100 Å to allow for wavelength differences from laser to laser and also to take account of wavelength shifts due to changes of laser operating temperature.

[iv] *Shot noise due to the detector dark current.* If thermal noise can be neglected and there is no background radiation the performance of a detector will usually be limited by the shot noise associated with its dark current I_d. The expression for the output signal-to-noise power ratio becomes

$$\frac{S}{N} = \frac{\eta^2 e P_s^2}{2h^2 \nu^2 B I_d} \tag{7.26}$$

and the criterion for a good detector is that it should have a high value of η^2/I_d.

A photomultiplier with an S-1 photocathode (see Section 7.6.2[ii]) has a dark current of about 10^{-12} A and a quantum efficiency at 9000 Å of 3×10^{-3}. A silicon photodiode may have a leakage current of 10^{-8} to 10^{-7} A and a quantum efficiency as high as 50 or 60%. Using η^2/I_d as a figure of merit it is seen that under dark-current limited conditions the performance of a silicon photodiode will be comparable to that of a photomultiplier. This comparison ignores differences in detector area and it must be remembered that P_s is the total signal power falling on the detector, so that if the photon flux is constant over the sensitive area of the detector a large device will give a correspondingly higher signal-to-noise ratio than a small one. However, in practical systems large detectors are not used simply to intercept more signal flux, but are necessary only to increase the beamwidth of the system or to simplify the optics, particularly if interference filters are used. In such cases the signal flux is rarely constant over the detector area and there is no general relationship between the detector area and the signal-to-noise ratio. For narrow beam systems large areas may be unnecessary, and the detector is often preceded by a small aperture stop to restrict the field of view and reduce unwanted background radiation. P_s and P_n are simply the total signal and noise powers passing through the aperture stop, and provided the detector is at least as large as the aperture its actual area is irrelevant.

[v] *Photon-limited conditions.* If all the above sources of noise can be eliminated the signal-to-noise ratio will be limited by the shot noise on the signal current itself. This condition represents the ultimate performance of a detector and is known as the photon-limit or quantum-limit. Eqn. (7.20) becomes

$$\frac{S}{N} = \frac{I_s}{2eB} = \frac{\eta P_s}{2h\nu B}. \tag{7.27}$$

This equation can be interpreted physically as implying that in order to detect a signal with a frequency response corresponding to a bandwidth B at least one photoelectron must be collected during each half-cycle.

Under photon-limited conditions the performance of a detector will depend only on its quantum efficiency, and it is unfortunate that the only class of detectors which have a low enough dark current and a high enough gain to be used at the photon-limit, namely photomultipliers, also have very low values of quantum efficiency.

[vi] *Amplifier noise.* The noise contributed by the amplifier immediately following the detector can be expressed as FkT/R_n, where F is the amplifier noise figure. This will depend on the gain, bandwidth and type of amplifier but with low-noise transistors may be in the region of 3 to 6 dB. R_n is the equivalent noise resistance of the amplifier and is often closely related to the input impedance which is the load resistance presented to the detector. Hence the effect of amplifier noise is very similar to that of the thermal noise considered in Section 7.6.1[ii] above, and it will increase as the input impedance is reduced to give larger bandwidths. Amplifier noise will become less important as the intrinsic current gain in the detector is increased and can usually be neglected if photomultipliers are used.

The performance of silicon photodiodes is often limited by the noise associated with the load resistance and the amplifier particularly at high frequencies. In this case resistive feedback amplifiers are often used since they can have an equivalent noise resistance much larger than the input impedance. For the basic circuit shown in Fig. 7.17a the equivalent noise resistance is equal to the feedback resistance R_F, whereas the input impedance is R_F/A where A is the voltage gain. The output voltage is simply the input current from the silicon photodiode multiplied by R_F. Since the equivalent noise resistance may be an order of magnitude greater than the input impedance, excess noise due to the amplifier can effectively be neglected and the output noise is merely the thermal noise calculated for the equivalent load resistance of the detector.

A specific amplifier of this type was described by Raines (1968) and the circuit is shown in Fig. 7.17b. It uses emitter follower stages at the input and output with overall resistive feedback, and is designed to work into 50 Ω output impedance. With this circuit noise measurements made over a range of bandwidths from 1 to 100 MHz (which can be obtained simply by changing the value of R_F and hence the input RC product) agreed very well with the theoretical figures for both thermal and shot-noise-limited conditions. In addition to giving an optimum detector performance the use of such an amplifier makes noise measurements on detectors easy to carry out and interpret.

7.6.2   *Comparison of detectors*

[i] *Choice of detectors for gallium arsenide laser systems.*   The construction and properties of the various types of photodetectors have been adequately described elsewhere and will not be discussed in any detail in

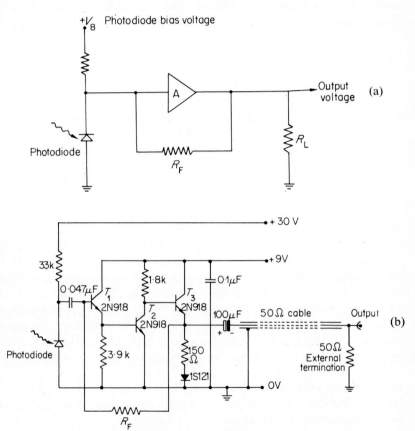

Fig. 7.17.   Resistive feedback amplifier: (a) Basic circuit.
(b) Practical transistor circuit

this chapter. A very good survey of detectors for use at high frequencies is given by Anderson (1966), and this includes a comprehensive list of references.

A comparison of the performances of the different types of detectors depends markedly on the bandwidth and background illumination of the particular system but some general conclusions can be drawn. In the visible region of the spectrum photomultipliers are best and the various types of travelling-wave or crossed-field electron multipliers can be used to give photon-limited performance up to microwave frequencies (Ross, 1966). In the infrared region beyond about 1 μ either germanium avalanche photo-diodes (Melchior, 1966) or some form of photoconductor (Sommers, 1966) must be used. However, at the wavelength of gallium arsenide lasers several types of detector may give similar performance, and the choice

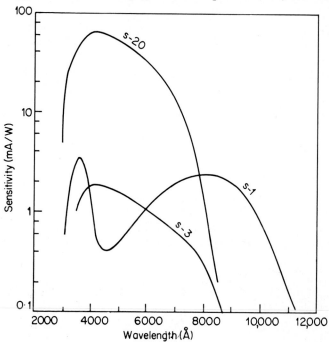

FIG. 7.18. Spectral sensitivities of photocathodes with long wavelength response

must be made by considering the detailed requirements of the system. The choice is usually between a photomultiplier, a silicon *p–i–n* photodiode and a silicon avalanche photodiode; these will be considered in turn in the following sections.

[ii] *Photomultipliers.* Fig. 7.18 shows the sensitivity of the three types of photocathode which have a long wavelength response extending beyond 8000 Å. It is seen that the S-1 photocathode is the most sensitive, but even this has a very low quantum efficiency, η, of only 3×10^{-3} electrons per photon at 9000 Å. It is sometimes claimed that selected S-20 photomulti-pliers can give better performance than S-1 tubes because they have a much smaller dark current, I_d. This may be true if the selection can be made from

an unlimited supply, but it is certainly not an economical way of improving system performance, and if necessary the dark current of the S-1 photocathode can be reduced a thousand times by cooling to about 200°K. At room temperature the dark current of an S-1 photomultiplier is about 10^{-12} A, and increases by a factor of four for each ten degree rise in

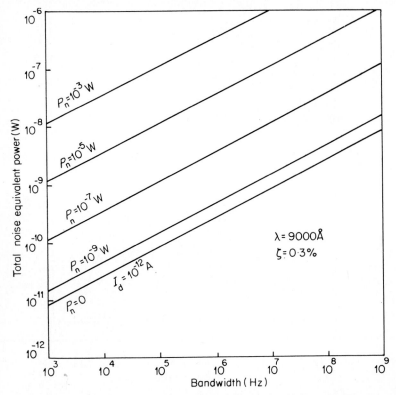

FIG. 7.19. Shot noise characteristics of an S-1 photomultiplier

temperature. It is the low dark current together with the virtually noise-free high internal gain that makes the photomultiplier a useful detector in spite of its low quantum efficiency.

Taking $\eta = 3 \times 10^{-3}$ and $I_d = 10^{-12}$ A the T.N.E.P. can be calculated as a function of bandwidth for different values of background power incident on the sensitive area using Eqns. (7.21) and (7.23). The results are shown in Fig. 7.19. In general the photomultiplier gain is sufficiently high for thermal noise and amplifier noise to be neglected and Fig. 7.19 enables the system performance to be predicted for a wide variety of conditions.

[iii] *Silicon p–i–n photodiodes.* A silicon photodiode can have a quantum efficiency as high as 60%, and with the correct choice of material resistivity and operating voltage the width of the depletion region can be controlled so that the spectral response peaks at about 9000 Å, as shown in

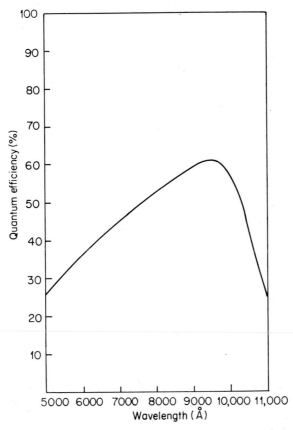

FIG. 7.20. Spectral response of a silicon *p–i–n* photodiode

Fig. 7.20. The dark current is about 10^{-8} A increasing by a factor of two for each ten degree rise above room temperature.

Using these figures the total noise equivalent power for the shot noise can be calculated, and the result is shown as a function of bandwidth and background power by the dotted lines in Fig. 7.21. However, as the *p–i–n* photodiode has no internal gain the thermal noise in the load resistance

11

will be important, and the T.N.E.P. corresponding to thermal noise is shown by the dashed line in Fig. 7.21. This has been calculated from Eqn. (7.22) assuming that the detector has a capacitance of 10pF (2 mm diameter), and that the optimum load resistance is chosen for each value of bandwidth, so that $R = 1/2\pi CB$. With pulsed operation the shot noise on

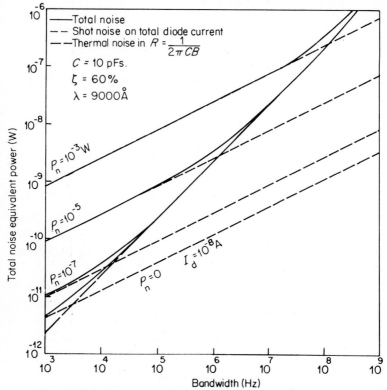

FIG. 7.21. Noise characteristics of a silicon *p–i–n* photodiode

the signal current can usually be ignored, and the detector performance is obtained by combining the shot and thermal T.N.E.P.'s to give the full lines in Fig. 7.21.

A comparison of Figs. 7.20 and 7.21 indicates that at 9000 Å a silicon photodiode can give a better performance than an *S*-1 photomultiplier for small bandwidths or with a high level of background illumination, but that in the dark or for very large bandwidths a photomultiplier should be used. The locus of points at which the performances are equal is plotted in Fig. 7.22, curve A, and this gives a general guide as to which detector is preferable under given conditions. However, in practice the choice is often biased towards the convenience of using a semiconductor device rather than a high-voltage vacuum tube. Slight changes in the values of η or I_d for the detectors will shift the curve in Fig. 7.22 appreciably, although the general conclusions remain valid.

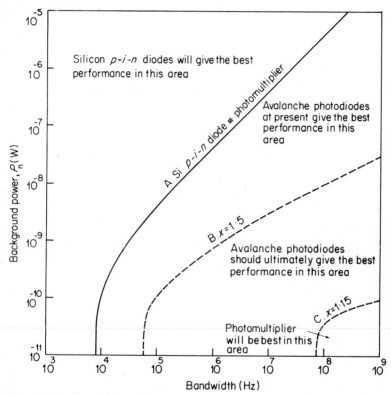

FIG. 7.22. Comparison between an S-1 photomultiplier and silicon photodiodes at 9000 Å

[iv] *Silicon avalanche photodiodes.* The effect of thermal noise on the performance of a silicon photodiode can be reduced by providing internal amplification, and the avalanche multiplication which gives rise to reverse breakdown can be used as a gain mechanism (Johnson, 1965). The avalanche photodiode is somewhat analogous to a photomultiplier, but since both electrons and holes are involved the amplification process is more complicated, and in fact the avalanche multiplication produces excess noise (McIntyre, 1966). If the signal current is multiplied by m the

signal power is multiplied by m^2, but the shot noise power is multiplied by m^{2x} where the value of x lies between 1·15 and 2 depending on the ratio of the ionization coefficients for electrons and holes (Biard, 1967). If no excess noise were produced the value of x would be unity, and for correctly designed silicon avalanche diodes it should certainly not exceed 1·5.

Hence, following Eqn. (7.22) the total noise equivalent power for an avalanche photodiode can be written

$$\text{T.N.E.P.} = \frac{hv}{\eta em} [2m^{2x} eBI_t + 8\pi kTCB^2]^{\frac{1}{2}}. \tag{7.28}$$

This is a minimum for an optimum value of multiplication given by

$$m_{\text{opt}} = \left[\frac{4\pi kTCB}{(x-1)eI_t}\right]^{\frac{1}{2x}}, \tag{7.29}$$

when the optimum performance is given by

$$(\text{T.N.E.P.})_{\text{opt}} = \frac{hv}{\eta e} [2eBI_t x]^{\frac{1}{2}} \left[\frac{4\pi kTCB}{(x-1)eI_t}\right]^{(x-1)/2x}. \tag{7.30}$$

The multiplication m is a function of the bias voltage applied to the diode, and in theory the optimum value of multiplication for a particular bandwidth can be obtained by simply adjusting this voltage. A tolerance of at least a volt will be permissible for values of m up to 100, but at bandwidths greater than 100 MHz where multiplications up to 1000 times may be required it will be necessary to stabilize the bias to within a few millivolts. The stability is required with respect to the diode breakdown voltage and it is possible to design suitable bias circuits using a Zener or control diode with a similar temperature coefficient of breakdown voltage to the avalanche diode.

If the development of avalanche photodiodes continues at its present rate devices will soon be available with the same values of quantum efficiency, active area, leakage current at low voltages and capacitance, as p–i–n photodiodes. With this assumption the T.N.E.P. can be calculated as a function of bandwidth and compared with that for an S-1 photomultiplier. This can be done for different values of x, and curves of equal performance can be obtained in the same way as for the p–i–n photodiode. Curves B and C in Fig. 7.22 are the results for $x = 1·5$ representing an excess noise performance that can easily be achieved, and $x = 1·15$ corresponding to the optimum design of a silicon diode. It is seen that the region of supremacy of the photomultiplier is considerably reduced, particularly for $x = 1·15$. Obviously if $x = 1$ could be achieved so that no excess noise was produced by the multiplication process the avalanche photodiode would give the better performance under all conditions of bandwidth and background noise. Results obtained so far with experimental avalanche photodiodes are very encouraging, and it should only be a question of time before the full performance anticipated by Fig. 7.22 is achieved in practice.

7.7 Gallium arsenide laser systems

7.7.1 Optimum system bandwidth

The underlying principles of active optical systems are extremely simple and have been treated generally by many authors (see, for example, Knight, 1959). The use of gallium arsenide lasers as the source of radiation in such systems has been discussed by Johnson (1963) and Vallese (1963). It is intended in the following sections merely to present some of the fundamental equations which illustrate the basic features of laser systems so that this chapter will serve as a convenient basis for the detailed consideration of any particular application.

The success of any optical system depends on achieving an adequate signal-to-noise ratio at the output of the detector. The determination of the noise level and methods of reducing it to obtain the maximum possible sensitivity were discussed in Section 7.6 above, and it only remains to calculate the amount of the transmitter power that is available at the detector. It is assumed that the laser transmitter emits pulses of radiation with a peak power P_t and a mean power P_m into a solid angle $\Omega_t = \theta_a \theta_e$, where θ_a and θ_e are the half-power points of the angular distribution in azimuth and elevation, respectively. P_t and P_m are related through the pulse width τ and the repetition rate f by the equation

$$P_m = P_t f\tau = P_t f/B, \qquad (7.31)$$

where B is the receiver bandwidth necessary to detect the pulses.

The pulse repetition rate, f, will be fixed by the particular application, and for high information rates the mean power, P_m, will be limited by the thermal dissipation of the laser. This means that the maximum energy in a single pulse is fixed and Eqn. (7.31) indicates that the peak pulse power can only be increased if the pulse width is reduced in proportion, so that $P_t \propto B$. The optimum pulse length depends on the dominant noise mechanism in the particular detector system. From Eqns. (7.24) and (7.26) in Section 7.6 it is seen that for background or dark-current limited conditions the signal-to-noise power ratio varies as P_t^2/B (since P_s, the peak power at the receiver, is proportional to P_t). Thus if $P_t \propto B$ the system performance increases with increasing bandwidth under these conditions. However, if the system is either thermal noise limited or photon-limited the signal-to-noise power ratio varies as $(P_s/B)^2$ or (P_s/B), respectively, so that the performance is independent of bandwidth and depends only on the energy in a single transmitter pulse. In general, then, it is worth-while reducing the laser pulse length in order to increase the peak power if the system is limited by the background illumination or the detector dark current, and if possible the bandwidth should be increased until either the thermal noise in the load resistance dominates or the photon-limited condition is reached.

Gallium arsenide laser systems are generally used for either communication or rangefinding, and these two types of application will be discussed in turn, but first it is necessary to consider the transmission of the laser radiation through the atmosphere.

7.7.2   Atmospheric transmission

The wavelength of gallium arsenide lasers is very near the visible region and will not penetrate fog or cloud so that gallium arsenide laser systems can only be used in clear conditions. Range gating can be used to improve the performance of rangefinders in haze or fog by eliminating the short-range backscatter and preventing the receiver from saturating. For communications the radiation only travels in one direction and the system will operate in somewhat worse conditions. Chatterton (1965) reported operation up to 1·8 miles when the visibility was only 0·6 miles.

Infrared transmission through the atmosphere has been extensively studied both theoretically and experimentally (Kruse, 1962; Bolle, 1965). At the wavelengths of gallium arsenide lasers the two main factors affecting the transmission are scattering by aerosols and particles, and the molecular absorption of water and carbon dioxide. Under normal conditions the largest effect is usually due to scattering by water droplets. This can be Rayleigh scattering if the droplet size is smaller than λ or Mie scattering for a droplet size comparable to λ. Scattering by solid particles can usually be neglected except at ranges above 10 km in very dry conditions. Absorption due to water vapour depends on the relative humidity and also varies markedly with wavelength. For a relative humidity of 60% about half the radiation will be absorbed in a distance of 10 km.

In general the transmission of the atmosphere can be expressed in terms of an attenuation or extinction coefficient, α, as

$$T_a = P/P_0 = \exp(-\alpha R), \qquad (7.32)$$

where R is the length of the transmission path. Relating measured values of α to the theories of scattering and absorption is virtually impossible because of the problem of accurately specifying the concentrations and sizes of all the relevant atmospheric constituents. In practice α is approximately inversely proportional to the visual range which is defined in terms of the visual contrast at 5000 Å (Kruse, 1962, p. 190). Numerical values of α vary between 0·05 km^{-1} for dry clear weather with a visibility of 40 km, and 30 km^{-1} for foggy conditions with a visibility of 100 metres.

7.7.3 Communications

For a simple communications system operating over a range R, the transmitter power intercepted by the receiver of area A_r is

$$P_s = \frac{P_t K A_r}{\theta_a \theta_e R^2} (\exp - \alpha R), \qquad (7.33)$$

where K is a factor representing the total optical losses in the system.

If α is small the range of the system will be proportional to the square root of the transmitter power. Eqn. (7.33) assumes that the transmitter beam more than fills the receiver aperture so that a better performance will be obtained if θ_a and θ_e can be reduced. However, apart from optical acquisition and alignment problems, it is difficult in practice to work with values of θ_a and θ_e less than about 1 milliradian because of scintillation effects due to temperature and pressure fluctuations in the atmosphere.

A compact single-channel voice communication system was described by Hannan (1967). The transmitter used four lasers operating at room temperature and focused into a single beam by four separate spherical lenses. The detector was a silicon photodiode which could normally be used without an optical filter as it was thermal noise limited. The range of the system was about 5 miles allowing 20 dB attenuation for poor atmospheric conditions. Much higher mean laser powers and pulse repetition rates can be obtained at 77°K, and Schiel (1965) reported communication over a 13 km path with a pulse rate of 1·125 MHz which allowed 24 pulse-code-modulated audio channels to be used. Rediker (1962) reported the transmission of a 4 MHz television channel by laser beam, but only over a short range in laboratory conditions.

7.7.4 Rangefinding

[i] *Range resolution.* The resolution of a pulsed optical ranging system will be approximately equal to the pulse rise-time in simple cases, although with sophisticated electronics the start of the pulse can be timed with a precision much better than the rise-time and without errors due to variations in pulse amplitude and system noise level (Kazel, 1963). Light

travels approximately 1 foot in 1 nsec, and a pulse with a 1 nsec rise-time will give a range resolution of about 6 inches since the time of flight corresponds to twice the range. Hence good resolution implies fast laser pulses and large detector bandwidths. High repetition rates are not necessarily required and the pulse length need be no longer than the rise-time, so that sufficient power can often be obtained from a laser operating at room temperature. Broom (1964), Goldstein (1967) and Koechner (1968) have described general ranging systems using gallium arsenide lasers, and Birbeck and Hambleton (1965) described a system specifically designed as an aircraft altimeter.

With simple pulsed systems a range resolution of a few inches can be achieved and this is usually sufficient. For higher accuracy it is possible to use microwave modulation superimposed on the pulse and employ phase-sensitive detection. With such techniques range resolutions in the region of 0·01 to 0·1 cm could be obtained in theory, but it would then be necessary to provide close control of the refractive index of the optical path.

[ii] *Ranging on a cooperative target.* For surveying purposes a cooperative target such as a corner-reflector can be used, and Eqn. (7.33) gives the amount of the transmitter power falling on a corner cube of area A_r. The diameter of the beam reflected back to the transmitter is

$$D = 2d + 2\Delta\theta R + \lambda R/d. \qquad (7.34)$$

The first term represents reflection from an effective diameter d, the second term allows for an error $\Delta\theta$ in the angles of the corner cube and the third term is the diffraction spread from the reflecting aperture for a wavelength λ. For ranges less than about 1 km the first term will dominate and a receiver of diameter greater than or equal to $2d$ will collect all the reflected power. A perfect corner cube will reflect the beam back exactly on the laser axis so that coaxial optics are needed for the transmitter and receiver. At short ranges the optics must be designed so that a large amount of reflected power is not obscured by the transmitter lens (see Fig. 7.15 in Section 7.5.2). At longer ranges the second and third terms in Eqn. (7.34) will dominate and a larger receiver will be required for optimum performance. If the reflected beam is larger than the receiver aperture the range will be proportional to the fourth root of the transmitter power, whereas at short ranges it is proportional to the square root of P_t. An imperfect corner cube gives six reflected beams spaced round a cone of semi-angle $\Delta\theta$, each of which has a diffraction-limited angular diameter of λ/d. At very long ranges it may only be possible to intercept one of these and coaxial transmitter and receiver optics may not give the best performance, particularly if the corner cube is very poor.

It may be thought that the very narrow beamwidth associated with the corner cube would give rise to very severe scintillation problems. However, the atmospheric perturbations which cause scintillation are much slower than the time of flight of the laser pulse so that the reflected beam traverses exactly the same path as the transmitted beam and any optical deviations are automatically cancelled out. The only scintillation observed will be that of the transmitter beam at the corner cube, and this will only be serious for very narrow transmitter beamwidths.

[iii] *Ranging on a non-cooperative target.* The received power in the general case will depend on the nature of the target and can be expressed in terms of a scattering function $F(\psi, \phi)$, where ψ and ϕ are the angles that the transmitter and receiver beams make with the target surface. This assumes a plane target and an integration has to be carried out for more complicated surfaces. Most naturally occurring targets have surface irregularities large compared with the laser wavelength and will scatter the radiation isotropically so that $F(\psi, \phi)$ is approximately constant for all values of ψ and ϕ. This value is called the scattering coefficient of the surface and will vary from zero for a perfect absorber to unity for a loss-less surface. For many artificial surfaces $F(\psi, \phi)$ has a specular as well as an isotropic component and an enhanced signal will occur at appropriate combinations of ψ and ϕ. The ratio of specular to diffuse scattering depends on the flatness and degree of polish of the surface, and it is usually necessary to make a direct measurement of $F(\psi, \phi)$ for the target in question.

If the target is large enough to fill the transmitter beam (as in the case of an altimeter pointing down at the ground) the total transmitter power P_t is available for scattering and the received power is

$$P_s = \frac{P_t K A_r F(\psi, \phi)}{2\pi R^2} \exp(-2\alpha R), \qquad (7.35)$$

where the symbols have the same meaning as in Eqn. (7.33). The factor 2π in the denominator implies that $F(\psi, \phi)$ is normalized by integration over a complete hemisphere. The radiation is now subject to atmospheric attenuation over a total path length of twice the range. If α is small it is seen that the range is proportional to the square root of the transmitter power.

If the target area A_t is smaller than the transmitter beam the power intercepted by the target can be found from Eqn. (7.33) and the power scattered back into the receiver is

$$P_s = \frac{P_t K A_t A_r F(\psi, \phi)}{2\pi \theta_a \theta_e R^4} \exp(-2\alpha R). \qquad (7.36)$$

In this case the range is proportional to the fourth root of the transmitter power.

Using the above expressions for received power and the equations in Section 7.6 for the noise the expected performance of any type of gallium arsenide laser system can be predicted, and it is comforting to close this chapter with the observation that in most cases the system performance in practice has agreed closely with that predicted by theory.

References

Adlam, J. H., and L. S. Holmes (1960). *J. Sci. Instr.*, **37**, 385.
Anderson, L. K., and B. J. McMurtry (1966). *Proc. I.E.E.E.*, **54**, 1335.
Ashburn, E. V., and R. G. Weldon (1956). *J. Opt. Soc. Am.*, **46**, 583.
Biard, J. R., and W. N. Shaunfield, Jr. (1967). *I.E.E.E. Trans. Electron. Devices*, **ED-14**, 233.
Biernson, G., and R. F. Lucy (1963). *Proc. I.E.E.E.*, **51**, 202.
Birbeck, F. E., and K. G. Hambleton (1965). *J. Sci. Instr.*, **42**, 541.
Birbeck, F. E. (1968). *J. Sci. Instr.* (Series 2), **1**, 788.
Bolle, H.-J. (1965). *Infrared Phys.*, **5**, 115.
Broom, R. F. (1964). Lasers and their applications. I.E.E. Conference, Sept. 1964.
Broom, R. F. (1965). *J. Sci. Instr.*, **42**, 123.
Brown, H. E., R. A. Bond and J. C. Bloomquist (1966). *Electronics*, **39**, Nov. 14th, 137.
Chatterton, E. J. (1965) *Proc. I.E.E.E.*, **53**, 2114.
Dousmanis, G. C., and H. E. Gross (1966). *Proc. I.E.E.E.*, **54**, 998.
Fenner, G. E. (1967). *Solid-State Electron.*, **10**, 753.
Goldstein, B. S., and G. F. Dalrymple (1967). *Proc. I.E.E.E.*, **55**, 181.
Hambleton, K. G., and F. E. Birbeck (1964). Lasers and their applications. I.E.E. Conference, Sept. 1964.
Hannan, W. J., J. Bordogna and D. Karlsons (1967). *RCA Rev.*, **XXVII**, 609.
Hansen, J. P., and W. A. Schmidt (1967). *Proc. I.E.E.E.*, **55**, 216.
Johnson, C. M. (1963). *Electron.*, **36**, Dec. 13th, 34.
Johnson, K. M. (1965). *I.E.E.E. Trans. Electron. Devices*, **ED-12**, 55.
Kapany, N. S. (1961). *J. Opt. Soc. Am.*, **51**, 32.
Kapany, N. S., and D. F. Capellaro (1961a). *J. Opt. Soc. Am.*, **51**, 23.
Karlsons, D., C. W. Reno and W. J. Hannan (1964). *Proc. I.E.E.E.*, **52**, 1354.
Kazel, S., and B. Ebstein (1963). *Proc. I.E.E.E.*, **51**, 1257.
Knight, K. V., (1959). *Proc. I.R.E.*, **47**, 1490.
Koechner, W. (1968). *I.E.E.E. Trans. Aerospace Electron. Systems*, **AES-4**, 81.
Kruse, P. W., L. D. McGlauchlin and R. B. McQuistan (1962). *Elements of Infrared Technology*, John Wiley and Sons, New York.
Matick, R. E. (1968). *Proc. I.E.E.E.*, **56**, 47.
McIntyre, R. J. (1966). *I.E.E.E. Trans. Electron. Devices*, **ED-13**, 164.
Melchior, H., and W. T. Lynch (1966). *I.E.E.E. Trans. Electron. Devices*, **ED-13**, 829.
Nelson, H. (1967). *Proc. I.E.E.E.*, **55**, 1415.
Prince, P. R. (1965). *Proc. I.E.E.E.*, **53**, 304.

Raines, J. A. (1968). Private communication.
Rediker, R. H., R. J. Keyes, T. M. Quist, M. J. Hudson, C. R. Grant and R. G. Burgess (1962). *Electron.*, **35**, Oct. 5th, 44.
Rosner, R. D. (1968). *Proc. I.E.E.E.*, **56**, 126.
Ross, M., R. B. Hankin, E. P. Dallafior and R. H. Swendsen (1966). *I.E.E.E. Trans. Aerospace and Electronic Systems*, **AES-2**, 62.
Ruzicka, J. (1958). In *Problems of Low Temperature Physics and Thermodynamics*, Pergamon Press, Oxford, pp. 323–329.
Schiel, E. J., E. C. Bullwinkel and R. B. Weimer (1965). *Proc. I.E.E.E.*, **53**, 2140.
Sommers, H. S. Jr., and E. K. Gatchell (1966). *Proc. I.E.E.E.*, **54**, 1553.
Sullivan, N. A. (1964). *Rev. Sci. Instr.*, **35**, 639.
Vallese, L. M. (1963). *Semiconductor Products*, Aug., p. 25.
Wilkinson, K. J. R. (1946). *J. Inst. Elect. Eng.*, **93**, Part IIIA, 1090.
Williamson, D. E. (1952). *J. Opt. Soc. Am.*, **42**, 712.
Witte, W. (1965). *Infrared Physics*, **5**, 179.

Author Index

Subject Index